Library of Congress Cataloging-in-Publication Data

Syon, Guillaume de, 1966–
    Science and technology in modern European life / Guillaume de Syon.
    p. cm. — (Greenwood Press "Daily life through history" series.
Science and technology in everyday life, ISSN 1080–4749)
    Includes bibliographical references and index.
    ISBN 978–0–313–33768–0 (alk. paper)
1. Science—Social aspects—Europe—History.   2. Science—Europe—History.
    I. Title.
    Q175.52.E85S96   2008
    509.4—dc22   2008026755

British Library Cataloguing in Publication Data is available.

Library of Congress Catalog Card Number: 2008026755
ISBN: 978–0–313–33768–0
ISSN: 1080–4749

First published in 2008

Greenwood Press, 88 Post Road West, Westport, CT 06881
An imprint of Greenwood Publishing Group, Inc.
www.greenwood.com

Printed in the United States of America

The paper used in this book complies with the
Permanent Paper Standard issued by the National
Information Standards Organization (Z39.48–1984).

10 9 8 7 6 5 4 3 2 1

For Corinne and Lucie, who discover daily new uses for old technologies

CONTENTS

PREFACE

This book is part of the Greenwood subseries "Science and Technology in Everyday Life," and thus seeks to offer a brief survey of how Europeans adjusted to the industrial age over the course of two centuries. In so doing, however, it also modifies traditional chronological approaches.

Modern European history is usually broken down into a "long" nineteenth century, stretching from the French Revolution in 1789 to the outbreak of World War I in 1914, and a "short" twentieth century, ending either in 1991 with the disintegration of the Soviet Union or even with the terrorist attacks of September 11, 2001. But discussing everyday life does not allow for so strict a delineation. As we shall see, many technological and scientific discoveries predate the modern era or do not affect most Europeans until decades later; the maturation process of these discoveries depended as much on engineers and scientists as it did on successful commercialization and the removal of social mores.

Geographically, Europe stretches from the Atlantic to the Urals, but the availability of published sources on which this book is based have focused mostly on Western Europe due in part to the accidents of history. Central and Eastern Europe in fact have as rich a history of scientific and technical discovery as anywhere else on the continent, but the focus of the work, namely, how people react and live with new machines and scientific applications, had to rely on the availability of records, many of which focus on the western part of the continent.

This work then offers a general reference on living with science and technology, but it is necessarily a snapshot of trends, habits, and reactions.

The selection of what seemed to have impacted most the lives of Europeans was difficult, for scientific and technological discoveries assume the existence of a seamless web, whereby one invention or discovery may affect everyday practices in a remote, yet proven fashion.

Because modern Europe was both the central theater of two world wars and the subject of new trends in postnational moves toward a global community, several chapters will emphasize the social and scientific debates that surrounded particular events such as the cholera, flu, and AIDS pandemics. A similar logic stands behind the inclusion of specific discussion topics in wider-themed chapters.

I would like to thank Kevin Downing at Greenwood Press, whose great patience was tested by multiple delays. Thanks also go to Tiffany Seisser, H. Brashear, and Amanda Foxcroft for their thorough copyediting work.

Traditionally, I should acknowledge scholars in the field. The incorporation of their work into the text (along with the sources cited in the bibliography) will make clear my debt to many historians, social scientists, and scientists. Since this work is intended for high school and beginning college audiences, let me instead mention students at Albright College, especially some intrepid undergraduates who enrolled in a prototype course on the history of technology in spring 2005, and the ones who took on other versions of said course in Albright's Accelerated Degree Completion Program since it was first offered in 2002. Their life experiences and insights from work helped refine my selection of examples for the narrative that follows.

INTRODUCTION: REVOLUTIONS

The advent of a new age in science and technology in Europe coincided with a dual economic and political revolution. The economic one, in the form of industrialization, found its roots in Great Britain in the eighteenth century, while the political one, in the form of the French Revolution of 1789, occasioned the rise of new ideologies and of nationalism in particular. In this context, science and technology came to be perceived as tools of state dominance, but also as supports for the improvement of life. Such notions, however, find their origins in a different kind of revolution that predates, then parallels the two mentioned: the scientific one.

This scientific revolution, which finds its origins in the seventeenth century, culminated in the Enlightenment, which emphasized rational thought as the solution to all social and political problems and sowed the seeds of the political and economic revolutions mentioned.

To understand the nature of the challenges Europeans faced, let us consider briefly the peculiar nature of this scientific revolution. Simply put, it became a new way of looking at the world. Though it has its origins in England, then the center of political upheaval (it is then that England became a constitutional monarchy, whereby the power of the king was checked by elected representatives), antecedents in the realm of ideas had existed in the preceding century, toward the end of the Renaissance period.

Then, for example, the model of the solar system had changed. By creating a new way of understanding how nature worked—and by solving long-standing problems in physics, astronomy, and anatomy—the theorists and experimenters of this period convinced their contemporaries that they had discovered new knowledge. They were not merely adapting the ideas of the famous figures of Greece and Rome. They went much further than those ancient scholars. Although their revolution began with a lot of questions and few clear answers, they ended up offering a promise of certain knowledge and truth that was eagerly embraced by their turbulent society.

Because it is primarily a revolution of the intellect, the best way to understand the importance of the scientific revolution is through the summary of a case study pertaining to astronomy and physics. In the early 1500s, traditional European ideas about the universe were still based primarily on the ideas of Aristotle, the great Greek philosopher of the fourth century B.C. These ideas had gradually been recovered during the Middle Ages and then brought into harmony with Christian doctrines by medieval theologians.

According to this revised Aristotelian view, a motionless earth was fixed at the center of the universe. Around it moved 10 separate, transparent crystal spheres. In the first eight spheres were embedded, in turn, the moon, the sun, the five known planets, and the fixed stars. Then followed two spheres added during the Middle Ages to account for slight changes in the positions of the stars over the centuries. Beyond the 10th sphere was heaven, with the throne of God and the souls of the saved. Angels kept the spheres moving in perfect circles.

Aristotle's views as revised by medieval philosophers also dominated thinking about physics and motion on earth. Aristotle had distinguished sharply between the world of the celestial spheres and that of the earth—the sublunar world. The spheres consisted of a perfect "essence," while the sublunar world (the earth) was made up of four imperfect, changeable elements. The so-called light elements—air and fire—naturally moved upward; the heavy elements—water and earth—naturally moved downward. Aristotle and his followers also believed that a uniform force moved an object at a constant speed and that the object would stop as soon as that force was removed.

Aristotle's ideas about astronomy and physics were accepted with minor revisions for two thousand years, and with good reason. First, they offered an understandable, common-sense explanation for what the eye actually saw. Second, Aristotle's science, as interpreted by Christian theologians, fit neatly with Christian doctrines. It established a home for God and a place for Christian souls. It put human beings at the center of the universe and made them the critical link in a great chain of being that stretched from the throne of God to the lowliest insect of earth. Thus, science was primarily a branch of theology, and it reinforced religious thought. So-called medieval scientists were hard to distinguish from theologians.

This desire to explain and thereby glorify God's handiwork actually led to the first great departure from the medieval, Aristotelian system. The departure was the work of the Polish clergyman and astronomer Nicolaus Copernicus (1473–1543). As a young man, Copernicus studied church law and astronomy in various European universities. He saw how professional astronomers were still dependent for their most accurate calculations on the work of Ptolemy, who had lived in the second century A.D. Ptolemy's rules allowed stargazers and astrologers to track the planets with great precision. Many people then and now believe in astrology—that the changing relationships between planets and stars influences and even determines the future.

Young Copernicus didn't like the way Ptolemy's rules worked, though. He preferred an old Greek idea being discussed in Renaissance Italy: that the sun rather than the earth was at the center of the universe. Never questioning the Aristotelian belief in crystal spheres or the idea that circular motion was most perfect and divine, Copernicus theorized that the stars and planets, including the earth, revolve around a fixed sun. Yet Copernicus was a cautious man. Fearing the ridicule of other astronomers, he did not publish his *On the Revolutions of the Heavenly Spheres* until 1543, the year of his death.

Copernicus's theory had enormous scientific and religious implications, many of which the conservative Copernicus did not anticipate. First, it put the stars at rest, their apparent nightly movement simply a result of the earth's rotation. Thus, it destroyed the main reason for believing in crystal spheres capable of moving the stars around the earth. Second, Copernicus's theory suggested a universe of staggering size. If in the course of a year the earth moved around the sun and yet the stars appeared to remain in the same place, then the universe was unthinkably large or even infinite. Finally, by characterizing the earth as just another planet, Copernicus destroyed the basic idea of Aristotelian physics—that the earthly world was quite different from the heavenly one. Where, then, was the realm of perfection? Where were heaven and the throne of God?

The Copernican theory quickly brought sharp attacks from religious leaders, especially Protestants. Hearing of Copernicus's work even before it was published, Martin Luther spoke of him as a new astrologer who wanted to prove that the earth moves and goes round. Luther quoted Holy Scripture: "so did Joshua bid the sun stand still and not the earth" (Joshua 10:10–15). John Calvin also condemned Copernicus, citing as evidence the first verse of Psalm 93: "The world also is established that it cannot be moved." Who, asked Calvin, would venture to place the authority of Copernicus above that of the Holy Spirit?

Catholic reaction was milder at first. The Catholic church had never believed in literal interpretations of the Bible, and not until 1616 did it officially declare the Copernican theory false. This slow reaction also reflected the slow progress of Copernicus's theory for many years.

Other events were almost as influential as Copernican theory in creating doubts about traditional astronomical ideas. In 1572, a new star appeared and shone very brightly for almost two years. The new star, which was actually a distant exploding star, made an enormous impression on people. It seemed to contradict the idea that the heavenly spheres were unchanging and therefore perfect. In 1577, a new comet suddenly moved through the sky, cutting a straight path across the supposedly impenetrable crystal spheres. It appeared time, to some exceptional people living in the sixteenth century, to radically change astronomy. One of the most important people to believe so was a Danish man named Tycho Brahe.

Tycho Brahe (1546–1600) had already established himself at the age of 26 as Europe's leading astronomer with his detailed observations of the new star of 1572. Brahe was an interesting man: After losing a piece of his nose in a duel and replacing it with gold and silver, he set himself up in the most sophisticated observatory of his day. There he trained a young assistant, Johannes Kepler (1571–1630), a German originally trained for the Lutheran ministry, who soon came to abandon his religiously based understanding of the universe.

Based on his work with all of Brahe's observations, Kepler formulated three famous laws of planetary motion between 1609 and 1619. First, building on Copernican theory, he demonstrated that the orbits of the planets around the sun are elliptical rather than circular. Second, he demonstrated that the planets do not move at a uniform speed in their orbits. Third, he showed that the time a planet takes to make its complete orbit is precisely related to its distance from the sun. Kepler's contribution was monumental. His work demolished the old system of Aristotle and Ptolemy, and in his third law, he came close to formulating the idea of universal gravitation.

While Kepler was unraveling planetary motion, a young Florentine named Galileo Galilei (1564–1642) was challenging all the old ideas about motion. Like so many early scientists, Galileo was a poor nobleman first marked for a religious career. However, he soon became fascinated by mathematics and ultimately made a great contribution to modern science: the elaboration and consolidation of the modern experimental method. Rather than speculate about what might or should happen, Galileo conducted controlled experiments to find out what actually did happen.

In the tradition of Brahe, Galileo also applied the experimental method to astronomy. His astronomical discoveries had a great impact on scientific development. On hearing of the invention of the telescope in Holland, Galileo made one for himself and trained it on the heavens. He quickly discovered the first four moons of Jupiter, which clearly suggested that Jupiter could not possibly be embedded in any impenetrable crystal sphere. This discovery provided new evidence for the Copernican theory, in which Galileo already believed.

Not surprisingly, the work of Galileo eventually aroused the ire of some theologians. After the publication in Italian of his widely read *Dialogue on the Two Chief Systems of the World* (1632), which too openly lampooned the traditional views of Aristotle and Ptolemy and defended those of Copernicus, Galileo was tried for heresy by the papal Inquisition. Imprisoned and threatened with torture, the aging Galileo recanted, "renouncing and cursing" his Copernican errors.

The accomplishments of Kepler, Galileo, and other scientists had taken effect by about 1640. The old astronomy and physics were in ruins, and several fundamental breakthroughs had been made. The new findings, however, had not been fused together in a new synthesis, a single explanatory system that would comprehend motion both on earth and in the skies. That synthesis, which prevailed until the twentieth century, was the work of Isaac Newton (1642–1727).

Newton was born into lower English gentry and attended Cambridge University. On one hand, he was a path-breaking scientist; on the other, he was intensely religious. He was also totally consumed by what he was doing. When writing the third book of the *Mathematical Principles of Natural Philosophy,* generally known as the *Principia,* Newton worked in his room for weeks on end, seldom leaving, even to eat. His goal was to bring together everything discovered by Copernicus, Kepler, and Galileo. Newton did this by means of a set of mathematical laws that explain motion and mechanics. These laws of dynamics are complex, and it took scientists and engineers 200 years to work out all their implications.

Nevertheless, the key feature of the Newtonian synthesis was the law of universal gravitation. According to this law, every body in the universe attracts every other body in the universe in a precise mathematical relationship: the force of attraction is proportional to the quantity of matter of the objects and inversely proportional to the square of the distance between them. With this law, Newton unified in one big system the whole universe—from Kepler's elliptical orbits to Galileo's experiments.

Considering such highlights of the scientific revolution, one wonders about their significance. First, the rise of modern science went hand in hand with the rise of a new and expanding social group—the scientific community. Members of this community were already linked together in the 1500s by common interests and organizations. Expansion of knowledge was the primary goal of this community, so science became competitive, and more advances were inevitable.

Second, the scientific revolution introduced not only new knowledge about nature but also a new and revolutionary way of obtaining such knowledge—the modern scientific method, which emphasized experiment though the use of new technical instruments (telescope, microscope, barometer, air pump, etc.). The air pump, for example, allowed for a better understanding of the properties of air pressure, burning, and breathing (Morton 13). In addition to being both theoretical and experimental, this method was highly critical, and it differed profoundly from the old way of getting knowledge about nature. It refused to base conclusions on tradition and established sources, on ancient authorities and sacred texts.

The scientific revolution had few consequences for economic life and the living standards of the masses until the late eighteenth century at the very earliest. True, improvements in the techniques of navigation facilitated overseas trade and helped enrich leading merchants. But science had relatively few practical economic applications. The close link between pure science and applied technology—a link that we take for granted today—did not exist before the nineteenth century. Thus, the scientific revolution was first and foremost an intellectual revolution. For more than 100 years, its greatest impact was on how people thought and believed. Yet, it paved the way for new orientations of the human mind.

Indeed, the stepchild of the scientific revolution was the Enlightenment. Whereas the scientific revolution had seen progress made specifically within the framework of astronomy, medicine, and other related sciences, the Enlightenment borrowed from its spirit of rational inquiry to expand it to social and political issues. Its impact can be seen not only in the framework of modern democracies, but also in the spread of scientific knowledge for lay educated people. This is reflected in publications as well as scientific associations.

The Enlightenment saw the production of a single work intended to show how knowledge could be useful: L'Encyclopédie, edited by Denis Diderot, was a Classified Dictionary of the Sciences, Arts, and Occupations. In other words, an inventory of knowledge from the most theoretical (definition of life) to the more run-of-the-mill (what is a key).

Over 20 years, between 1752 and 1772, some 28 volumes were published. This was not an impartial endeavor. Criticisms and partisanship abounded in the entries, and from the third volume onward, the volumes were censored and published illegally (and sold to private subscribers). The flawed entries are obvious, even by the standards of the eighteenth century. Yet, what Diderot sought to achieve, namely, a kind of revolution in the minds of men, hoping to free them from prejudice, while not exclusively the realm of his work, contributed to the spirit of the Enlightenment.

As the educated gentry was exposed to new scientific and technical ideas, it sought to gather in groups to discuss such discoveries. Salons—essentially gathering in someone's house for the purposes of socializing and exchanging ideas based on what one read—also served as places to learn of scientific experiments. At such gatherings, scientists would offer paid lectures on so-called natural philosophy. The first such lectures dated back to 1705 and were offered not so much as training for future physicians (though some were) but as a form of rational entertainment that allowed the happy few to test their understanding of the newly developing sciences (Morton 21). Most importantly, however, such openness to science through public lectures does not mean that science offered anything other than entertainment. To many, the notion of progress associated with scientific and technological endeavors in the nineteenth and twentieth centuries simply did not exist: One got interested in science for the sake of knowledge and little else.

This is further confirmed in the establishment of scientific academies. These were different from the state-controlled academies (such as the Royal Society in England) set up in the seventeenth century. Scientific academies flourished all over Europe and welcomed anyone who had what was called "curiosity"—that is, a willingness to learn about science. Societies gathered, listened to learned papers, and tried to hold conferences. They also had a "cabinet of curiosities," basically a room with all kinds of natural artifacts: deformed animal bones, rocks, and a number of oddities the owner of said cabinet might find amusing.

Women were eventually allowed into such gatherings, but as spectators rather than active members. The novelty of science meant that few saw any threat to a female presence. As for children, several popular science books began to appear in the eighteenth century focusing on the lives of inventors (Morton 54). In many ways, the tradition of heroic inventors that they introduced in the United Kingdom and continental Europe remains present in popular publications to this day.

THE BEGINNINGS OF THE INDUSTRIAL REVOLUTION

Aside from a revolution of the mind, another with far-reaching physical consequences also finds its origins in the eighteenth century. The Industrial Revolution started in England and eventually spread to France after the French Revolution and then to Germany and Italy. It barely affected Eastern Europe until the late nineteenth century. What we need to ascertain is why this was the case.

The reasons for this are complex and are in fact still subject to ferocious arguments among not only historians but also anthropologists, sociologists, and, of course, economists. We need only look at the agreed upon facts.

By the eighteenth century, Europe had accumulated a huge amount of wealth. This means that it had improved its agriculture but that its trading system was also more efficient, allowing merchants to accumulate benefits and reinvest them. Furthermore, states competing with each other encouraged new manufacturing methods, with an eye to acquiring more efficient and cheaper weaponry. Increases in commercial transactions also meant more tax income, which in turn could finance military endeavors. The best part of all this is that the European system could swallow all this because it had the manpower to do so.

In Western Europe, in contrast to other continents and nations, the population was growing, and governments no longer had the means to channel that manpower (through serviage, for example) and had to rely on market forces. In other words, people found jobs instead of being forced into them, simply because of the size of the population. Between 1800 and

1900, the size of the European population almost doubled, reaching some 460 million. The next revolution, an agricultural one, would limit the risk of disease by reducing famine, though it would not eradicate it.

The agricultural revolution, often considered the first green revolution in modern history, changed the modes of food production. By the eighteenth century, a new pattern of farming emerged, one in which food production was done for the market, not simply for personal/family needs. Obviously, traditional patterns of interaction among peasants and lords remained, but farmers were also drawn to the market simply because it was a chance to make a profit, especially for those who were free land owners. New techniques of farming were introduced (see chapter 1), and the primary scene of such change was in England.

Still, such factors do not explain why England jumped into the Industrial Revolution alone. When comparing the country to its French neighbor, the latter featured an increasing population, a rich intellectual and cultural life, and an even greater estimated gross national product. In the interest of maintaining economic and military advantages, the French government had also taken to founding early technical and trade schools. The reasons for England jumping ahead are a mix of political factors (England was a constitutional monarchy; France was an absolutist regime), economic ones (the taxation system favored the nobility and the clergy while squeezing the small middle class and the large peasantry), and social practices (even small-scale production relied on hand labor, with no place for machinery of any kind).

England, on the other hand, featured a labor pool of farmers who could no longer earn a living on the land and needed income. It also offered easy access to open-air coal mines and enjoyed good communication links (especially waterways). Finally, the middle class was politically connected, and the laws were modernized to allow people to switch professions and guilds. Their increased wealth, for which they could also thank limited engagement in war (Landes 46), rather than being stashed away was reinvested. This placing of funds to allow the manufacture of cheaper goods in response to demand reflected not only the flexibility of the British system, but also the ability to respond to market demand for specific goods (Landes 52). In light of this, what technologies made the changes possible?

The market in which the Industrial Revolution was first noticeable was that of cotton, for which there was a very strong demand. Cotton is durable, washable, versatile, and cheaper than linen cloth or wool. Thus there was a great incentive to produce cotton goods cheaply. Slavery in the American colonies afforded this opportunity. Unfortunately, cotton-processing centers in England could not meet the demand. It was based on the *putting-out* system. Putting out is an aspect of the preindustrial era, whereby rural families devoted more time to industrial work, primarily weaving, spinning, or finishing textiles. This was an aspect of cities' influence on the countryside. Merchants concentrated their activities in the city proper, yet relied on countryside labor to complete products. This actually familiarized inhabitants of the countryside with the manufacturing process and spread a little wealth (though not for long). The problem for merchants was that they were limited to using the supply of labor in their own districts. To fix this, they would go farther and farther into the countryside, but then they faced problems of communication and of regular supplying and quality control (someone closer to the city was more easily accountable).

Things began to change in the 1730s with the invention of the fly shuttle, which permitted construction of larger and larger handlooms. This, however, meant that extra yarns of cloth were required, a supply that was still missing, thus resulting in a bottleneck. In response to this, several inventors tried to build a system in factories that would allow for quicker production of the yarn. One of the first was Richard Arkwright (1732–1792). He used water power to spin cotton fibers into wheels and then twist them into yarn. This worked well, but it required setting up shop near a river. Enter James Watt (1736–1819), who invented and perfected a working steam engine (1785), which Arkwright soon adopted.

The next phase shows clearly how technology works in a kind of seamless web. Here, socioeconomic implications began to enter the fray. Indeed, the British now had as much yarn as deemed necessary, but there were too few handloom weavers to meet the demand! Edmund Vartwright's perfecting of a power-driven loom would eventually resolve that problem. Things did not go smoothly, however. Not only did the machine require perfecting, but weavers' opposition was strong. Indeed, once perfected, it was successful among factory owners who viewed it as a great means to watch over workers' output and to hire fewer of them: one person (often a child) could watch over two power looms with outputs up to 15 times greater than that of a skilled handloom weaver.

Consequently, cotton textile production jumped tenfold between 1760 and 1785 and another tenfold between 1785 and 1825. The British population around that time had been 80 percent occupied on farms. By 1900, the number was down to 25 percent.

The incredible dynamism associated with the eighteenth century should not, however, be mistaken for any substantial advance in the realm of technology. Muscle and wind power remained the main determinants of substantial work. All experience in this realm was based on everyday observation and transmission of knowledge orally in workshops. As one historian put it, there was "no appreciation of thermal cycles, no coherent concept of energy, no science of thermodynamics, no understanding of metabolism" (Smil, *Creating the Twentieth Century* 14). All this would come in the course of two centuries to follow. In so doing, it would also change family lives faster than in the entire preceding millennium. To understand this, let us first consider the changes in the countryside and the city overall.

CHRONOLOGY

<table>
<tr><td>1789</td><td>Beginning of the French Revolution, which inaugurates the long nineteenth century in European history.</td></tr>
<tr><td>1823</td><td>British medical journal The Lancet is founded.</td></tr>
<tr><td>1831</td><td>Justus von Liebig and several others discover chloroform.

Michael Faraday discovers electromagnetic induction, which suggests that mechanical power could generate electricity.</td></tr>
<tr><td>1832</td><td>A cholera epidemic begun the previous year spreads across Europe. Several other waves will follow in 1847, 1853, and the 1860s.</td></tr>
<tr><td>1845</td><td>The rotating press is invented, which revolutionizes the print and newspaper industries.</td></tr>
<tr><td>1846</td><td>Le Verrier surmises the distance of planet Neptune through mathematical calculation.</td></tr>
<tr><td>1847</td><td>First use of chloroform as an anesthetic in Great Britain.</td></tr>
<tr><td>1851</td><td>The Great London Exhibition at Crystal Palace opens on May Day.</td></tr>
<tr><td>1854</td><td>John Snow (1813–1858) tracks the origins of a cholera epidemic in London to a water pump on broad street.</td></tr>
<tr><td>1855</td><td>Henry Bessemer patents the process that bears his name for making stronger and cheaper steel.</td></tr>
</table>

1859   Battle of Solferino is the largest land battle since the time of Napoleon. It influences Henri Dunant (1828–1910) to create the International Committee of the Red Cross.

Charles Darwin publishes *On the Origin of Species*.

1867   Alfred Nobel discovers dynamite.

1877   Charles Gros and Thomas Edison invent in parallel the phonograph.

1888   Inauguration of the Pasteur Institute, one of the first biomedical research centers.

1889   The Eiffel tower is constructed for the Paris Universal Exposition.

1895   Wilhelm Conrad Röntgen discovers X-rays.

First tires installed on automobiles.

1897   The diesel engine makes its first appearance.

1899   The German Bayer Company introduces powdered aspirin.

Sigmund Freud (1856–1939) publishes *The Interpretation of Dreams*, a seminal work in the history of psychoanalysis.

1900   Max Planck establishes quantum theory.

First flight of a rigid Zeppelin airship.

1902   The first electric kettle is introduced in Great Britain. A model with automatic safety shut-off, however, won't appear until 1929.

1903   First flight of the Wright brothers in the United States.

1905   Einstein writes three fundamental papers, all in a few months. The first paper claims that light must sometimes behave like a stream of particles with discrete energies, "quanta." The second paper offers an experimental test for the theory of heat. The third paper addresses a central puzzle for physicists of the day—the connection between electromagnetic theory and ordinary motion—and solves it using the principle of relativity.

The first diesel engines for ships are produced by the Swiss company Sulzer.

Carl Linde, Germany, patents the first ammonia compression refrigerator.

1906   Bakelite, a synthetic resin, becomes the first plastic material for sale in the United States and soon becomes available in Europe.

1907   Germany sets up the first automatic telephone exchange.

1908 Appearance of the world's first cable car at Wetterhorn, near Grindelwald (Switzerland).

The first automatic phone connector is installed in Hildesheim, Germany.

Jacques Brandenberger, in Zurich, Switzerland, patents cellophane.

Fritz Haber, Germany, synthesizes ammonia, thereby paving the way for the industrial production of ammonium nitrates, necessary to agriculture, the following year.

The Wright brothers demonstrate their mastery of the air during presentations in France.

1909 Louis Blériot flies across the English Channel.

1910 First publication of vitamin charts and definition of blood types.

1911 Lord Rutherford defines the first representation of the atom.

1914 In Basel, Switzerland, the CIBA pharmaceutical company produces the first hormonal preparations.

The first radio concert is organized in honor of the queen of Belgium.

World War I begins.

1915 Gas is used for the first time against Allied troops.

1918 The Spanish flu pandemic causes an estimated 25 to 50 million deaths worldwide.

The Frigidaire company commercializes the first mass-produced refrigerator under the name "Kelvinator."

World War I ends.

1920 The British Marconi company initiates the first series of radio programs.

1921 The Eiffel tower becomes the central sender of radio programs in Paris. In Lyon, the first private and public radio stations are set up.

Discovery and use of the first tuberculosis vaccine, the BCG.

1926 Erik Rotheim, Norway, invents the first aerosol.

1927 First tetanus vaccine.

1928 Alexander Fleming isolates penicillin, an antibacterial agent.

1937 Destruction of the *Hindenburg* airship at Lakehurst, New Jersey, ends the passenger rigid airship era.

1939 Otto Hahn and Lise Meitner discover nuclear fission and chain reactions.

First use of DDT insecticide to combat Colorado beetle outbreak in Switzerland.

World War II begins.

1943 Chemist Albert Hoffman, working in Switzerland, accidentally discovers LSD.

1946 Streptomycin, discovered two years earlier by Selman Waksman (1888–1973) as the first antibiotic effective against tuberculosis, is now tested in Great Britain.

1954 First atomic power station to produce electricity for industrial use begins operations at Oblinks, USSR.

1956 First civilian nuclear power plant in Great Britain.

The SECAM color television system is patented. It goes into service in the 1960s and evolves into the European standard PAL-SECAM.

1959 Switzerland becomes the first nation globally to completely automatize its phone network.

1960 The Rome Olympic games are the first broadcast though the *Eurovision* consortium of television networks.

The contraceptive pill is introduced.

1966 The last human-operated telephone connection is closed down, ending an automatization process that took 60 years.

1969 Maiden flight of the Franco-British supersonic transport Concorde.

1972 First flight of the European Airbus A 300 B, first model of the Airbus consortium.

1973 The first oil shock disrupts traffic in Western Europe.

1974 The information chip is patented. It eventually becomes a standard safety feature on many European-issued credit cards.

1977 The first microwave ovens appear in the Western world.

The last Orient Express train from Paris to Istanbul completes its run.

1978 In the Netherlands, the Philips corporation unveils the first compact discs (CDs).

1979 The nuclear accident at Three Mile Island in Pennsylvania brings new awareness of the danger of power plants in Europe.

1981 First cases of AIDS reported to the Centers for Disease Control in the United States.

1986   The Amstrad computer company (United Kingdom) standardizes the PC format for the European market.

1988   RU486, a medication causing spontaneous abortion, is invented.

1990   After centuries of planning and failed attempts, the tunnel under the English Channel is completed.

1995   Appearance of the World Wide Web, first developed in 1989 at the European Center for Nuclear Research (CERN).

2000   The American global positioning system (GPS), once reserved through "selective availability," is no longer restricted. GPS units are now sold across the globe.

2003   The Concorde supersonic airliner is withdrawn from use.

2005   Skype, a free phone software program, becomes compatible with most European systems and allows voice communication over the Internet.

       First flight of the Airbus A380, an aircraft bigger than the legendary Boeing 747.

2007   A modified French high-speed train (TGV) establishes a new world speed record, reaching over 356 mph.

2008   The rise in the price of oil prompts a reconsideration of consumerism and threatens to downsize the global economy substantially.

I

AGRICULTURE: FROM FARM TO TOWN

It is not possible to account for all agricultural transformation across Europe, for factors as varied as industrialization, geography, population increase, and political events fashioned the agricultural world into a varied canvas. The irony becomes clearer considering that local farmers, tied to land by economics and tradition, began producing in the modern era for consumers far away from the region. A corollary of this economic shift involves technology, or lack thereof. Agriculture experienced substantial changes in the early and modern eras, but less so at the technical than at the economic level afterward. By the time mechanization appeared in the twentieth century, the massive impact of the pauperization of farmers and their transformation into members of the working class was largely complete.

Before this happened, life in the countryside remained largely unchanged, as exemplified in the tools farmers preferred to use. Most tillage (preparing the land for crops) was done with wooden tools, though occasionally a bit of iron might appear thanks to the local blacksmith. Oxen and horses pulled the plow, under control of a plowman. In Europe plows usually had a mould-board, which buried the seed as the earth was turned. This was exhausting work that called for the plowman to use upper-body strength to maintain the proper direction of the plow, all the while pushing down against any root or rock the blade might encounter. Wheeled plows were preferable, especially in the heavy soils made muddy by fall rain, but tradition accounted for the slow adoption of these tools (Yarwood 48). One noticeable change came in the late eighteenth century when the first cast-iron plows made in factories were sold. Other farmers, however, preferred to use the dibble method, which consisted of drilling holes in the soil at regular intervals, and the farmer's wife and children would follow behind to drop seeds in the holes. Overall, the labor-intensive nature of farming continued despite the fact that improvements were available.

The first noticeable difference between the city and the countryside involves the implementation of scientific and technical discoveries. In the eighteenth and nineteenth centuries, hundreds of dissertations appeared that offered suggestions for improvements in grain output, animal husbandry, and farming tools. Most of these experiments, carried out in theory but not in practice, quickly disappeared into archives (Van Bath 239). There were, of course some exceptions.

England saw the most innovations in agriculture in the eighteenth century. The primary reason for this is the concentration of land in greater properties that the English nobility and gentry managed. Agricultural experts also traveled to the Netherlands to study the farming methods there, as they were reputed to be the most efficient. However, they did not simply copy what they saw but sought to offer further innovations. These included the introduction of new crops that could feed cattle, as well as new forms of animal husbandry that strengthened existing stock. Such innovations reflected experimental approaches, but technology, too, saw notable improvements.

Jethro Tull (1674–1741), for example, made several important discoveries in the eighteenth century. In particular, he suggested that the yield of crops could be raised through row planting and mechanical seeding. He was instrumental in recommending horses be bred for use in place of oxen, and he offered some redesigns of farming tools, notably the plow. He did oppose the use of fertilizers, thinking plants had enough access to sustenance in the soil. He was incorrect, but his primary concern was to avoid fertilizing with horse manure, which contained weed seeds.

Tull's inventions were sometimes ignored for years, even though he published several of his recommendations in a book, *The New Horse-Houghing Husbandry: An Essay on the Principles of Tillage and Nutrition*. Tull's theories demonstrate the difference between agricultural and industrial output. When he offered his suggestions in the 1730s, cereal prices were at an all-time low, which meant there was little incentive to invest in improvements. The primacy of economic factors over technical ones meant that it would not be until the nineteenth century that Tull's theories gained a following.

The other reason for the slow adoption of technology in rural areas relates directly to the farmers themselves. Agriculture is a trade usually learned from a young age and weaved together with cultural traditions and social mores, notably an important sense of filial piety. To change things was at times seen as disrespectful of the forefathers' legacy. As one historian noted, many stuck unadvisedly to old ways, thus compromising their economic well-being, but little could be done to change this state of affairs until the 1850s.

By the mid-nineteenth century the advent of new large-scale transport technologies in the form of railroad carriages and large steel-hulled ships affected grain prices throughout Europe and prompted agrarian lobbies to demand that tariff barriers be raised to ensure protection of their production. While in the short term this may have been an attractive solution to a falling price crisis, it actually contributed in the long run to inflation, as people had to pay more for basic food staples, which meant salaries had to be raised, too (Trebilcock 50–51).

Furthermore, industrial demand for certain crops prompted a shift in the production of grain and vegetables. The linen industry expected more flax; madder and woad were in heavy demand in the dye industry. All these were exacerbated by the increase in consumer spending: The increased demand for beer, associated with increases in leisure time, would prompt more production of barley and hops for beer, while tobacco farmers faced pressures to deliver more of their product for industrial manufacturing of smoke products.

These industrial scale demands combined with horticulture in some countries. Whereas flower growing had been a matter of private interest for centuries, the increased demand of the middle class prompted full shifts in some realms. All of this required either more labor or, alternately, mechanization. In England in particular, the adoption of threshing machines in the 1820s and 1830s cut costs while increasing productivity. As wealthy landowners adopted the machines, however, they also put hired hands out of work. The resulting famine among certain groups of the farming population caused riots, and in some farming areas, displaced workers attacked the portable threshers, smashing them. To give their opponents the illusion that the movement was widespread (it in fact had limited support among small landowners who could not afford the new machines), the rioters would sign with graffiti that included the name of Captain Swing. There was no such person. Yet, the myth grew. It took two years to quell the riots completely. Displaced by the machine, those who were not arrested migrated to cities. Some starved to death. The protesters were not so much Luddites as they were simply hopeful for employment, however. The fact that threshing machines used fewer men was what bothered them. Though the popular challenge to machinery eventually subsided, others, in the form of scientific improvement, came about.

BRINGING SCIENCE INTO THE COUNTRYSIDE: COWS, VACCINES, AND FERTILIZER

One central facet of agricultural modernization came to include how much milk a cow could produce. Where agriculture was diversified, cows were put to pasture. However, in the premodern era, they did not produce milk year-round. This was due to the folk notion that food became blood when consumed and, in the case of cows, eventually turned to milk. It took the work of Justus von Liebig in 1842 (*Animal Chemistry*) to start setting the record straight. In addition, agricultural experts realized that it might be preferable to actually separate permanently animal husbandry from arable land to introduce a new cycle of permanent milk production. A farmer might do both, but he needed to keep them separate so that they could support each other effectively (Orland 171).

This process of *animalization*, whereby cattle dung would be used to fertilize arable land, which itself would have some fodder plants in some fields (clover or sainfoin), took time to implant itself. The economic imperative, combined with the appearance of an international grain market and the existence of more railroad links, convinced farmers in continental Europe to follow suit. There, a new profession, *Wiesenbaumeister* (master of greenland) developed with an eye to expanding existing farmsteads and building artificial meadows. Some farmers, eager to add animal husbandry to their trade, sought existing stock, notably from Switzerland and the Netherlands (Orland 174).

But cows had to be "learned," not just brought to a new field. Farmers did not understand the science of animal husbandry, and many felt that their own cows were better than the imported ones because they were used to local soil. The best way, it turned out, to change their minds was not to enforce instructions but to institute competition with cash prizes. This is what happened in Bern, Switzerland, in 1806: Cattle exhibitions were born, and the practice of studding slowly followed. By 1848, when Switzerland gained a formal constitution for the first time in its history, standardized Swiss cattle shown at international exhibitions were deemed part of Swiss identity, along with its milk, butter, and cheese production. In a country traditionally poor in natural resources then, the dairy cow seemed a godsend. But to gain a foothold into this trade, farmers needed to be schooled, and herdbooks became the central element of classification of their stock, which included such things as a point scale and had to account for individual variations.

This concept of individual variation, though seemingly obvious, was a novelty in the nineteenth century where the new science of statistics suggested there had to be a perfect outcome if all instructions were followed. Charles Darwin's work *The Variations of Animals and Plants Under Domestication* (1868) codified what many breeders had discovered, namely, that two cows raised in the same herd and under similar conditions might not produce the same amount of milk because of genetic inheritance. Thus, science in this case joined learned practice (Orland 182).

In Denmark, farmers agreed to form herding societies with inspectors who would record cow performance and test the milk to advise on the removal of the lowest-producing head. By World War I, dairy societies had successfully taken over farming interests. Any farmer who wished to practice animal husbandry would have to follow scientific practices if he wished for his trade to thrive. Unwittingly, a process of competition had, within a century, allowed science to penetrate one of the most tradition-oriented realms of human activity. Another example, this time in response to a health crisis, would also have repercussions.

Pasteurization

Though the name of Pasteur is associated with the method used to preserve milk products from dangerous contamination, the French scientist was involved in the discovery of other scientific solutions. As a chemist and a friend of the French emperor Napoleon III, Pasteur set out in 1864 at the request of the latter to investigate the diseases afflicting grapes. The incidence of spoiled production was such that it threatened the entire wine industry in France. Pasteur set up a laboratory near a vineyard at Arbois in 1864, where he studied samples and their interaction with various crystallized acids. Pasteur came to the conclusion that microorganisms were responsible for the spoiling of wine in distilleries. He also found, through experimentation, that the microorganisms could be destroyed by heating the wine for a few minutes at 131°F then cooling it down suddenly. This allowed for a longer conservation of the product and the destruction of microorganisms within it: Pasteurization was born. It would take the scientist several more years to refine his understanding of the process, which he first applied to beer with the publication of a manual for beer brewers in 1875, and later to milk.

Encouraged by Pasteur's success in discovering a cure for wine disease, the French government, through Senator Jean-Baptiste Dumas (who had previously mentored the young scientist in Paris) also asked Pasteur to study silkworm disease. The silk industry in France competed directly with Asian production, yet an illness plagued silkworms. Pasteur's investigation showed that infectious agents were indeed the cause and were transmitted both through contagion and the hereditary principle, meaning that future generations could carry the disease. Building on previous knowledge he had acquired in the study of fermentation, Pasteur then argued that specific microbes caused illnesses and that they were in fact foreign bodies attacking the organism. Pasteur then suggested an industrial process to rid the silkworm of the illness. Though he is of course credited with saving the industry, it is important to consider that the authorities in charge of implementing the cure were just as heroic in convincing silkworm producers to accept the scientific process. Necessity likely played a role, as it did in the third and perhaps most important scientific improvement to farming.

In the nineteenth century, it became clear that crop rotation would not be sufficient to respond to a notable growth in human population and demand for agricultural products. Scientists who had studied the Irish famine, where the potato crop failure had wiped out hundred of thousands and forced as many as a million to emigrate, were convinced this could happen again. What was required, they argued, was not only diversification, but increase in yields. Barring that, the world might starve. Increasing crop yields involves using better fertilizers. Potassium, phosphorous, and nitrogen all fit the bill and had been identified in the 1840s, when the modern study of plants and soils became an established field in industrial and academic circles. The problem, however, was finding enough nitrates with high nitrogen content. Guano (bird droppings) could sometimes be found in quantities, especially in Latin American coastal areas: Chile saltpeter, for example, accounted for over 60 percent of all world reserves in the nineteenth century, but any supplies were quickly bought out. The challenge for industry, then, became to find a way to fix nitrogen into a compound that could be spread onto fields. Some in existence were useful, but not very efficient overall. This is where German-born Fritz Haber's contribution becomes all the more important.

The son of a prosperous chemical merchant, Haber soon decided to turn to an academic career, a daring move at a time when most junior academics survived on family money, which was lacking in Haber's case. His intensive early studies were in electrochemistry and thermodynamics and soon gained him the position of professor of physical chemistry at the Technische Hochschule, Karlsruhe (1898). His most important work, which began in 1904, was the synthesis of ammonia from hydrogen and nitrogen, a process described in the Enlightenment, yet one that had not been carried out successfully. By 1908, Haber was able to convert nitrogen from the atmosphere into liquid ammonia (NH_3), the raw material for making nitrogen fertilizer (N_2). What was still needed was the means to do it so industrially.

Fritz Haber and Karl Bosch developed the process jointly, and their effort was an important example of cooperation between industrial and academic circles. While the reaction between nitrogen gas and hydrogen gas to produce ammonia gas had been known for many years, the yields were very small, and the reaction very slow. Haber and Bosch and their coworkers determined the conditions necessary (high temperatures and very high pressures) and the catalysts necessary (a variety were found, the cheapest and most effective being oxides of iron with traces of oxides of other common elements). German industry also developed the high-pressure equipment necessary to run the process. By 1913 a chemical plant was operating in Germany, producing ammonia by means of the Haber-Bosch process. Without the Haber-Bosch process, as it became known, the green revolution of the early twentieth century would not have taken place. The Haber-Bosch process relied among other things on the ability to control high-pressure gases in the generation process, something Germany was able to master well into World War I. In fact, it would not be until the 1920s that the process would be exported to the rest of Europe and to the United States. Nowadays, nitrogen farming is essential, though the pollution levels of rivers and even seas have raised alarm among environmentalists as well as international organizations. Still, it is clear that the green revolution of post–World War II was carried out in part thanks to the discovery and mass production of ammonium nitrate (Smil 2001).

At the close of the "long nineteenth century," farming had already undergone substantial transformation due to the Industrial Revolution. The penetration of technology into the countryside, however, remained limited. Traditionally, a patriarchal system continued to dominate, and traditions, including the adoption of some machinery into festivals, suggested a sense of stability. World War I changed this, especially in countries where farm workers were called up. The epic destruction of life took a heavy toll on some regions, to the point where entire male contingents of rural families disappeared. Nowadays, village monuments to the dead of the war bear silent witness to the tragedy: Fathers, sons, uncles, and cousins all appear together in the name lists. The impact on farming becomes clearer if one considers that widows had to take over. This was less of a social shock to the women involved than it was in the urban realm. Women in the countryside did as much work, including physically demanding work, as men did. This was the first time, however, that in many cases no men were around to take over, let alone offer advice. Financially, this sometimes delayed the decision to acquire tractors. In places such as the United Kingdom, where massive government effort had prompted the replacement of horses (requisitioned for the war effort) with vehicles the Ford companies sent in from overseas, the point was moot. In France and Germany, though, the arrival of the tractor in the field would have to wait, even though other technologies were already present in rural areas.

Farming Technologies

Although some farming technologies received a strong welcome, other machines emanating from urban realms were far less accepted. The automobile, for example, did not fare well in the first two decades of its existence in the countryside. The lack of traffic rules was partly to blame, as was the fact that roads tended to serve as spillover areas for roosters and other farm animals in Europe, but the country resisted and occasionally resorted to violence. Outside Berlin in 1912, cases of several drivers decapitated by metal wires strung between trees were reported. The animosity rarely reached such levels but was easy to understand. When a driver came through, he often traveled far too fast to stop in time, and the squashed hen was a loss for the farmer. Drivers rarely stopped (automobiles were notoriously difficult to restart), and when they did, it was usually to ask for assistance with towing their car. Many an artist caricatured the clash. What the American countryside nicknamed "the devil in God's country" had its counterpart in Europe. Nonetheless, the horse's days were numbered as new machines came about.

Technological improvements to farming are usually associated with the twentieth century, aside from the early tractors. Yet, some activities had already started in the nineteenth century. Grain threshing, for example, had not changed since the Middle Ages and required most members of the family to help process the wheat crop by using a flail. Despite the increased number of hands, the slowness of the work was noticeable, and it was very hard physically. Mechanical threshing using crankshafts appeared by the 1850s, with many such models having been invented by small farm owners who were not interested in raising production so much as remedying the limited labor available in remote areas. Many such machines were even presented at juried competitions, and what surprised the judges was the fact that manual labor was not excluded but made part of the machine.

Even in areas where village or district administrations might have bought steam-powered machines, grain threshing was often kept until late fall or winter, to keep workers employed. Nevertheless, local administrations sometimes rented a steam-powered thresher, or even offered special financial rewards to local inventors of such machines. The

One of the culture shocks that the rural areas of Europe experienced was the appearance of the automobile and its impact on traffic. Narrow roads became extremely dangerous, and the lack of traffic rules, as implied in this caricature, meant the horseless carriage still had a long way to go before becoming part of the countryside. Source: Astra, Geneva, Switzerland; reproduced by permission.

rationale that administrators used to suggest that machines should be used pertained of course to efficiency: More grain would be threshed mechanically, whereas that done by hand often left some grain intact. One machine owner reportedly kept his machine on display as a warning to workers that they could be displaced if they failed to do their threshing work properly (Weber 188).

In some areas, social resistance was not a matter of tradition but of topography. Some crops needed the scythe, which caused a substantial loss of grain because the terrain would not accommodate a mechanical or steam thresher. Other areas still called for the sickle because, although slower than the scythe, the former offered a more uniform cut with more grain preserved. True, shifting the way one worked in the field would have likely solved the problem and prompted a switch to mechanical means of harvesting, but the practices learned and applied for centuries were just too hard to shed in a matter of months (Weber 124).

The mechanical thresher also had other unintended consequences besides putting people out of work. Most farm roofs in the early nineteenth century still were made of thatch. This was of course a fire hazard, but many farmers, even those who could have afforded it, refused to switch to slate or tile. The arrival of the mechanical thresher forced the change where administrators could not: Threshers broke the straw into such small pieces as to make it unusable for roofs. The victims in this case, though, were the poorest peasants who could not afford tile (Weber 17).

By 1914, the mechanization of the countryside was far from complete, but there was nevertheless a greater awareness of how the machine could serve the purposes of farmers.

Though the advent of the internal combustion engine in the late nineteenth century would have likely impacted the countryside around cities, it was another event that precipitated change. World War I imposed a state of total war on the economy by subordinating all societal needs and practices to those of the front. In so doing, it caused a dearth of horses in the countryside and took away many able young men. It also created a food shortage that could only be remedied through the introduction of mechanization. The role of the state in this respect is substantial. While authorities would indeed play a role in encouraging the use of modern machinery, they would also remain involved to ensure that supply, pricing, and steady farming did not fail. The food shortages of the Great War remained deeply ingrained in memory, and this was one of its impacts: greater mechanization. Yet, other machines would precede the tractor.

In some areas, many farmers grew frustrated with the physical exhaustion that hand-operated mechanical threshers caused and asked their co-ops to rent or buy steam-powered ones. If anything, some noted that the steam thresher had become akin to the proverbial office water cooler and more. Not only was its noise (the whistle of steam and clattering of parts) the signal of social gatherings, but the appearance of a rented machine in

Fig. 349. — *Batteuse Damey à manège direct placé sous la batteuse.*

An 1881 model of a thresher. The horses operate the mechanism, a notable improvement over the more common use of manual labor. Such a setup, while reducing the number of farmhands, nonetheless remained a part of socialization as celebrations involving decorating the machines and preparing food took place. Source: *Dictionnaire des arts industriels,* 1881.

the field also reflected the crowning of a year's work on the crop. Hence, a festive mood was often associated with it, and extra meals with copious offerings were made available to all who brought the cuts to the threshing machine. Often then, the new mechanical contraption came across as a kind of visitor to the village or field where it was set up.

One of the new tools was the milking machine, which cut down on labor and time, allowing farmers greater output. The machine itself was a novelty (and would not become widespread until after World War II in most of Europe), but a few appeared in the interwar years. Reminiscing on his time spent in the English farmland in the 1930s, farmer Walter Spreadbury noted:

> When old Cave [a new owner] came here first after taking over the farm from old Joey Nicholls . . . he brought along with him a Hosier milking machine. Now this machine was the first one anybody had ever seen or heard of round this way. Now this meant that those old cows didn't have to be taken twice a day to the yard for to be milked, because this was going to be done out on them downs with this machine, and on top of that, it only needed one cowman and one boy for to milk the whole of them instead of about four of them. (Spreadbury 6)

The twentieth century also saw technological changes in the farmland that had little to do with farming, such as the purchase of freezers. First acquired by farmers' co-ops to help store extra meat prior to shipment, they soon made their way into rural households' kitchens or cellars. In France, this brought changes in cooking practices, as more food was prepared in advance, partly as a way to plan more economically. The freezer, however, also encouraged more gifts of food, as it allowed people to keep what could not be eaten right away. This tradition of stocking up might have replaced the proverbial sticking of the pig in preparation for winter, when sausage and hams supplemented a potato diet, but in rural areas culturally bent on tradition, it simply added to the availability of choice, and the porcine member of the farmhold still met its fate.

As the other main implement in the kitchen, the stove did not gain as quick an acceptance. In some cases, mistrust continued to prevail. The speed of the gas oven upset rural users. It was not uncommon to encounter in farmers' kitchens up until the late twentieth century a gas oven alongside a wood oven. The latter's heat was deemed a "smoother heater" that allowed the farmer's wife to slowly cook or bake, which gave her time to go and milk the cows before coming back to check on her recipes.

THE END OF THE HORSE AGE

Surveying the technology of farming in Europe requires that we look at what it displaced. Like in the United States, the civilization of the horse came to an end in Europe, only to remain as a vestigial element associated with sports and occasionally farming. This is also why mobility in the twentieth century became a true revolution: It removed traditions going back several thousand years. Though the horse was central to the countryside, it was as much a fixture in the city. For the sake of clarity, its disappearance in both realms is chronicled here.

The decline of the horse age had begun in the nineteenth century. Though actively used in industry, the spread of the railroad had already placed horse couriers in financial difficulty. The early automobile had not been much of a threat to the horse's central place in society, and the two means of travel were even deemed complementary.

On January 11, 1913, bystanders along the Saint-Sulpice-La-Villette horse carriage line witnessed an odd funeral: a horse drawn carriage (known as an *omnibus*) with travelers and also covered with funerary crowns bearing the inscription *RIP*. The last trip was far merrier than any real funeral and reminded some of an anniversary celebration. On the bright side, one would not have to worry about stepping into horse dung when crossing over the street rails. In addition, the replacement gas-powered buses and electric trolleys appeared faster, even though the smell of gasoline from buses seemed just as unpleasant to riders.

In many ways, this symbolic end (occasional horse-drawn trolleys and cars could still be found in the interwar years) had not been foreseen. Far from diminishing the use of horses, the train had instead rerouted its use.

In the city, it remained king, as delivery services and public merchants relied on horse-drawn carriages. Pedestrians also faced the risk of horrible accidents (Pierre Curie, husband of physicist Marie Curie, perished under a carriage wheel in 1906), but such risks were deemed part of the urban experience.

Even though the car did eventually displace the horse from the city, its use in the countryside remained strong. Farmers considered their beasts not only part of their financial capital but also of their identity as farmers: Horses served every purpose, and while putting one down brought some financial relief (by the nineteenth century, restaurants bought horse meat), it was also a difficult sentimental matter. Farmers taught both sons and daughters the technique of farming using bull- but also horse-drawn plowshares. Small farms survived thanks to their animal husbandry. World War I would change this.

The mobilization of European armies also prompted a requisition of all horses from the countryside for cavalry as well as artillery-drawing purposes. Consequently, bigger farms found that it would be necessary to switch to tractors. While mechanization did increase in the interwar years, the selling off of horses varied from one country to another. Great Britain was among the fastest to go "horseless," while France and Southern European nations were among the slowest. Regardless of the rate of decline, by 1950, Europe was in effect a horseless society. Horses had become so rare and expensive that farmers had no choice in many cases but to switch to tractors.

The tractor was shown to reduce physical effort and increase productivity, but the disappearance of the horse from urban and rural areas had far more economic implications than one can imagine. The reduced need for farm workers encouraged a new wave of urban migration and also changed everyday life in the countryside. The new "beast" required maintenance and refueling, but this did not have to be done on a daily basis. Schedules closer to those of the city were now possible.

Many professions disappeared as a result of the introduction of tractors and cars. The leather market crashed because harnesses were no longer needed. Staples such as oats and corn faced overproduction because horses no longer existed to eat them. Great Britain had approximately 20 million horses in 1920; even railroad companies there at the time continued to have large stables, which, in some cases, had almost as many horses as there were locomotives.

Indeed, some countries kept horses as valued agrarian tools well into the twentieth century. In Finland, horses remained in great demand because of their use in logging. Yet, much of this changed by the 1960s. In France, the term *traction horse* to refer to beasts pulling heavy loads was deleted from administrative lists in the 1970s: Too few were still in use (for, say, clearing brush and fallen trees) to warrant maintaining the category. In some Eastern European nations such as Poland, the horse remained in use until the 1980s but fell rapidly after the shift to democracy: Privatization prompted many to seek modern machinery.

Consequently, the horse as a tool has all but disappeared. Farmers on the island of Knossos, Greece, are known to capture wild horses to help stomp the grain in the summer, and some farmers prefer using them to gather crops that tractors might crush. Some armies (such as Switzerland's) even maintained horse-and-donkey units for mountain operations where mechanized travel was impossible well into the 1980s, but the reality of modern warfare prompted their withdrawal, too. Thus, the technology that the horse represented has shifted to being an art, one devoted to leisure. True, horse racing and breeding involves substantial scientific preparation, but the interaction between horses and the public has become limited. Instead, other modes of transportation became part of everyday life in twentieth-century Europe.

AFTER WORLD WAR II

The process of emptying the countryside became complete in the decades after World War II. Though farming remained an activity that states protected with tariffs and sponsored economically, it was in fact in decline. In France, the very low level of mechanization on farms yielded a fully mechanized farming industry within three decades. The farming population, fully one-third of France (12 million out of 40 million), has fallen to half a million nowadays. The paradox of this transformation is that it happened less through machinery than through a social network. In this case, the Catholic Agricultural Youth (JAC), which encouraged raising the educational level of its members, handed out brochures and booklets and undertook the organization of travel to trade shows where they were introduced to mechanized milking machines, tractors, and, of course, the huge threshing tractors imported from the United States. Although the JAC eventually lost its influence, it was replaced by other organizations with similar agendas concerning modernization as well as cooperation and joining of farms to ensure more efficient use of the expensive tractors, and so forth. The French government got involved, too.

Beginning in 1946, French authorities tried to reenergize agriculture by setting up a series of experimental farming districts. The endeavor failed to convince local farmers to emulate what they saw, and in 1952, a new wave of pilot projects was initiated with the intent of increasing productivity and thus the income of small and medium-sized French farms. Throughout the 1950s, a series of pilot farms were established, involving both technocrats from the agricultural ministry and trained farmers. In so doing, such pilot farms cleared the way for the expansion of transitional grassland, ensilage, and dairy farming. The corollary of such success, however, was that it changed the culture of the land. Instead of learning their trade from their elders, young farmers required advanced schooling, and the high cost of the equipment associated with such reforms accelerated the merging of individual farms into cooperatives. It also created a gap of knowledge between farmers who remained wed to older farming methods and those who had acquired the new, technically oriented ones.

The reaction of farmers varied considerably. On the one hand, many welcomed the new methods that lightened the workload and increased yields. In parallel, added expectations of governmental assistance created unexpected tensions. For example, cattle fever in the French region of Cantal in the 1950s prompted the French government to offer vaccines for sale. The farmers, however, expected the vaccination of the stock to be free, and few consequently chose to resort to it. Though most understood clearly the need for protection, they also cast the scientific element of animal husbandry squarely on the back of the government. This state of affairs would also be reflected, away from science and technology, in vested interests of various political coalitions throughout Europe that sought to maintain tariffs to help local farmland.

The availability of industrially prepared food for animals also had an impact on farming. Producers now had to budget cattle feed, among other purchases, whereas they used to be able to count on the availability of hay in the harvested fields. Though the latter remains true for small farm

In the late twentieth century, collecting hay to feed cattle is no longer a community undertaking. Where entire families would once assemble for several days to complete the task, two operators and their tractors can do the same task in this mountainous farming area in a matter of hours. Source: Astra, Geneva, Switzerland, reproduced by permission.

businesses, the need to supplement cattle diets remains, to ensure either greater milk production or the required weight to bring a beast to the slaughterhouse. The advent of mad cow disease changed this.

Whereas animal illnesses had largely been mastered by the mid-twentieth century (including rabies, anthrax, and several other fevers), mad cow disease changed farming attitudes in Europe and created a conundrum. First identified in Great Britain in 1986, mad cow disease, known scientifically as BSE (bovine spongiform encephalopathy), caused an embargo on British meat three years later. Authorities forbade farmers from using cattle feed that contained animal products. Indeed, it was found that processed carcasses of sheep had been crushed and mixed in as powder into such feeds. Bovines are vegetarian. The reaction of farmers and of the population was disbelief and then panic. In 1993, for example, the first case of transmission of BSE to humans, in this case a British farmer, resulted in the patient's death from Creutzfeldt-Jakob disease.

CONCLUSION

The nineteenth century witnessed the end of the self-sustaining farming family, as it moved away from the countryside, or into the concentration of specific production, be it corn, wheat, or animal husbandry. Furthermore, the impact of technology also made agricultural work into a more individual endeavor, which changed the traditional relationship villages enjoyed. The migration begun under conditions of pauperization was accelerated through war and the attractiveness of urban work conditions. Once almost 40 percent of the population on average, the mean of farming families has fallen below 5 percent in most of industrialized Europe. Rare exceptions do exist, of course, mostly in the rural areas of southern and central Europe. There, mechanization is slowly making headway. What mechanization allowed for was the increase of farmland exploitation, while applied scientific methods increased yields. A European farmer used to produce for himself and later for his immediate surroundings. Cooperative units and wide-scale food production changed this dimension; the farmer now produces for a continent.

2

BUILDING THE CITY AND ITS URBAN CULTURE

Between 1750 and World War I, Europe industrialized in no fewer than three waves: between 1780 and 1820, then between 1840 and 1970, and finally in the decade preceding the Great War. This latter wave is sometimes considered part of the second one, only interrupted by an economic depression and then accelerated in part through the arms race that preceded World War I. The Industrial Revolution originated in Great Britain, where certain technological innovations including the steam engine and new textile spinning methods cleared production bottlenecks. The combination of technological know-how, large-scale production, and structural transformation of society was first realized there, while on the European continent, it encountered difficulty in several regions due to such factors as the French Revolution, political crises, and war. Belgium was the first to successfully emulate the British, while other countries had regional success based on political priorities. It is important to note that the required technological knowledge to achieve industrialization evolved. Whereas the first wave made successful use of British canals (water communication was the swiftest in the late eighteenth century) as well as cotton spinning and iron puddling, a half century later, the railroad mattered in transport, and engineering went hand in hand with steel mills (Trebilcock 42).

Urbanization accompanied all three waves, and the proportion of populations living in cities grew as societies were transformed. At the end of the eighteenth century, the overwhelming majority of all populations throughout the world lived outside of cities. In Great Britain, the proportion of the population living in urban areas was 25 percent in 1831, then exceeded

50 percent two decades later, and broke 77 percent by 1901. In Prussia, and then Germany, the urban transition lasted longer, beginning with 26 percent of urban migration in 1816 and exceeding 50 percent only in 1900. France, on the other hand, would not reach 50 percent until the interwar years.

THE FIRST WAVE'S SOCIAL IMPACT

From the very first wave on, a new social strata appeared: the working class. In order to get a handle on what this theoretical development of a class system meant for most people in real life, let's look at the impact of the industrialization process in Europe. The social disruption caused by technological invention and its application in an economic setting, though in the long run positive, caused substantial difficulties.

Factory workers did try to protest such a state of affairs, and their rebellion, early on, was against the machine. In the winter of 1811–1812, for example, glove makers in Nottingham, England, rose up against their employer by smashing the stocking frames they needed to do their jobs. The machine breaking in fact deprived them of work, but they stuck to the notion that they needed to rise against the anonymous form their labor had assumed. Their leader, Ned Ludd, may have been fictitious. They assumed the name of Luddites and called for a return to a preindustrial order. While this longing was unrealistic, the term became associated with any aversion to technology. The machine was here to stay, and the order it brought with it was difficult to survive.

Factory workers on the whole earned too little to sustain a family, and thus the employment of women and children became as necessary to survival as it was advantageous to employers, who appreciated their greater dexterity and the lower wages they would accept. The largest factories were the cotton mills, where commonly half of the laborers were women and a quarter were children. A class was thereby formed of men, women, and children, clustered about a source of employment, dependent on cash for their subsistence, and subjected to the rigid discipline of the factory. Awakened before dawn by the factory bell, they tramped to work, where the pace of production was relentless and the dangers from machinery and irate foremen were great.

Hours in the factories were long, 14 a day or occasionally more. Discipline was such that if your attention lapsed at all during a workday of 14 hours or more, even for stopping to help a neighbor, you would be fined and reprimanded. Children were frequently beaten (Cunningham 87–89, 121–33), and because they could pass in narrow areas, they were often made to do dangerous and growth-stilting work around machines. As one witness testified:

> Factory labor is a species of work, in some respects singularly unfitted for children. Cooped up in a heated atmosphere, debarred the

necessary exercise, remaining in one position for a series of hours, one set or system of muscles alone called into activity, it cannot be wondered at—that its effects are injurious to the physical growth of a child. Where the bony system is still imperfect, the vertical position it is compelled to retain, influences its direction; the spinal column bends beneath the weight of the head, bulges out laterally, or is dragged forward by the weight of the parts composing the chest, the pelvis yields beneath the opposing pressure downwards, and the resistance given by the thigh-bones; its capacity is lessened, sometimes more and sometimes less; the legs curve, and the whole body loses height, in consequence of this general yielding and bending of its parts. (Gaskell)

Beyond the horrors of factory life, working-class members of industrializing societies suffered greatly in their homes as well.

The Industrial Revolution changed the landscape of the city. As paupers moved in closer to factories and sought dwellings, new urban inconveniences appeared. Crowded areas with no infrastructure quickly turned squalid: to come home from work meant leaving one hell, associated with the machine, for another. The safe haven associated with the home did not exist for the vast majority of working Europeans.

Despite the overpopulation, cities took steps to ensure a more stable development of their urban structure. This of course was done primarily in well-to-do areas rather than in working-class communities, where speculators tried to pile as many dwellings together as they could (Yarwood 96). Railways added to the havoc, as large swaths of buildings turned to rubble to make way for tracks.

Major city redesigns also took place using newly available technologies. The most famous example was that of Paris where Baron Haussmann led the redrawing of the French capital by laying large boulevards and digging a new sewer system. The project took years to accomplish and displaced thousands of workers, who were forced to seek new dwellings rapidly. The conditions wherever they went were horrible.

The housing in factory areas was hastily constructed and barely large enough for large families that moved with their many children who also worked in the factories. Such basic services as garbage pickup and water supply did not exist. Sewage was a serious health problem. In the mid-nineteenth century, Louis Villermé, a military surgeon turned social investigator, toured a working-class living quarter in Lille, a major cotton-manufacturing town in northern France, and described conditions that characterized one-third of Liverpool as well as some of Lille's citizens:

The poorest live in the cellars and attics. These cellars . . . open onto the streets or courtyards, and one enters them by a stairway which is very often at once the door and the window. . . . In their obscure

cellars, in their rooms, which one would take for cellars, the air is
never renewed, it is infected; the walls are plastered with garbage . . .
If a bed exists, it is a few dirty, greasy planks; it is damp and putres-
cent straw; it is a coarse cloth whose color and fabric are hidden by
a layer of grime; it is a blanket that resembles a sieve . . . everywhere
are piles of garbage, of ashes, of debris from vegetables picked up
from the streets . . . of animal nests of all sorts; thus the air is unbreath-
able. . . . And the poor themselves, what are they like in the middle
of such a slum? Their clothing is in shreds, without substance, consumed, covered, no less than their hair, which knows no comb, with
dust from the workshops. (Villermé 82)

More than anything else, family life was considered by contemporaries to be the greatest casualty of the industrialization process. From
the viewpoint of the aristocracy or the new middle class, the slums
represented centers of immoral behavior and misery where children
abounded and parents paid no heed. Much of this stemmed from contemporary beliefs that women working outside the home represented
the end of a caring and nurturing family life as it had existed from
the beginning of time. Although women in the preindustrial age (e.g.,
putting-out system) had certainly worked long hours, the fact that they
now left the home struck many middle-class commentators as disas-
trous for the family structure. Furthermore, as factories ceased hiring
families and children together, women now rarely worked side-by-side
with their children; parents and children now headed off to different
factories in the mornings.

While, in fact, working-class parents and their children had often been
separated in the past by the different tasks of the putting-out system, they
themselves believed the strain on the family was found especially in the
lack of housing and the conditions of work. The women, especially, would
work 12–14 hour days at the factory and then return home to prepare the
meal, care for the children, and maintain the home, which was generally
an urban shack much too small for the large family.

The squalor did not go unnoticed. Many social reformers, whether
socialists or religious, became concerned that technology in the form of
factory work and urbanization was actually destroying the family unit
among the poor. To prove their case, they claimed that increases in prostitution and illegitimacy (children born out of wedlock) were on the rise.
While it is true that almost 50 percent of all births were out of wedlock in
some industrial cities, a true measure of social breakdown would have to
be considered in terms of child abandonment. These did occur, of course,
but the numbers rose only when economic conditions worsened. In other
words, so long as adult and teenage members of the family had employment, it was highly unlikely that the family fabric would completely
whither away.

However, child labor remained the central problem for all concerned reformers. In the nineteenth century, children under 16 were active in the cotton industry and in mining especially. Their youth made them extremely useful (and cheaper) over the course of half a century when some technological solutions had yet to be introduced to fix cotton-spinning machines. The complexity of early cotton machines was such that factory managers realized quickly that adults seeking to convert to factory labor were unable to learn the new trade, whereas children developed a quicker and longer-term understanding of how the machines worked. These children often remained on staff into adulthood, which meant that fewer children were needed a generation later. In fact, in England, the number of children employed in cotton mills started to fall ahead of the child labor legislation imposed in the 1830s (Galbi 357). However, it is important to realize that work conditions remained very harsh, with children expected to help clean a machine or to help unjam it by crawling in small spaces under it. It was not uncommon to see children get scalped as the flying shuttle slid by and caught their hair, or for them to either break or crush a limb as a heavy part of the machine folded over. Some were wounded or killed when they fell asleep, exhausted, while monitoring the machine's progress. Their numbers remained substantial, despite the risk, and because it was often the only access to income they could find. In France in the 1850s, 12 percent of the cotton workers were children under 16.

Child labor in mines was a different story. Mines were inherently more dangerous than cotton mills for children, though their small physical size enabled them to work a variety of odd jobs. Some were used as so-called trappers to help move along coal carts. Whenever they heard one sliding along on its rails, they would pull a rope to open a trapdoor (they were in a small enclosure) that would let the cart pass. The closing of the trapdoor prevented the cart from occasionally sliding back and was also a safety measure against drafts that might help ignite flammable gas. This work, though relatively easy, was often described as lonely: Unlike mining teams where sometimes entire families would work together, a trapper spent his day listening for the cart and did not see anyone until it was time to go back up. Other jobs involved actual coal extraction. Because entire families would work in the mine, either jointly (for salary reasons) or on alternate shifts (so that some might survive in case of accident), children were sometimes involved in extracting coal from veins too narrow for adult miners to reach. Others would carry loads in buckets on their back, either to a cart or all the way to the elevator area. Children who experienced such work, when reaching adulthood, would already be in poor health and unable to be as productive (and therefore earn more). Reformers' pressures succeeded in banning children under 13 from work in the mine through the Miners Act of 1842.

In other nations, however, such legislation came more slowly and was not always enforced properly. Though social reformers operated from a sound principle of concern for childhood, they often failed to consider the reasons for child labor, namely, that it was the only way for a family to make ends meet. By limiting the income earned through children, they thus restricted economic well-being without offering an alternative. It would take the social reform of several governments, over several decades, before some of the horrors associated with the Industrial Revolution's punishing technological requirements subsided. For example, though the Socialist movement of the nineteenth century had made some progress in obtaining better working conditions for workers, and pressures from middle-class family groups had helped ban children's labor, typical factory work remained difficult (60 to 70 hours), partly due to the need to travel long distances to work. Here is how a Swiss skilled metal worker put it in 1912:

> The day went something like this. In the winter, rise around 4:15 A.M., then leave the house by 4:50, run 4 kilometers [about 2.2 miles] to the train station. After the train ride, walk another half-hour to the office. The day schedule was 6:30 to noon and 1:30 to 6 P.M. Then, the same exhausting ride home. (Treichler 59)

Although this man's account suggests an element of normality in his life, it is devoid of comfort. The beginnings of the next wave of industrialization would remedy some of the difficulties the working class encountered, but not as much as social reformers had hoped.

The next wave of industrialization is often called the second Industrial Revolution and consisted of two waves, interrupted by the economic depression of the 1870s and 1880s. However, it is in that era that new technologies and applied sciences appeared. These included chemical industries, electricity, bicycles, and the early automobile. Characteristic of this wave was the diversification of industry. No longer was the steel mill of the railroad the central technological feature of industrialization, but rather a series of supply industries, from transmission gear for subway cars to tires for bikes.

By the 1880s, then, one would think that all industrialization was done on a mass scale. This was not the case. Economic and technological historians of the Industrial Revolution found that not only did different paces of industrialization continue, thus resulting in a regional adaptation of industrial conditions, but in some cases, technology even contributed to helping keep smaller production units alive, such as in Italy, where electricity became the key to such survival (Trebilcock 49).

In the case of steel production, a new method, known as the Bessemer process, was introduced in the 1860s to help form higher quality steel and iron by increasing the temperature inside the smelter furnaces (this allowed removal of impurities from molten iron, which means that you could use lower grades of iron ore to make steel: one could purify that iron more easily to obtain similar qualities of steel). The quality of the steel meant it could be incorporated more easily into construction, notably building cities.

Adoption of steel, however, was slow, except where the railroad was involved. Construction techniques were well established. And it was not until buildings grew higher or new technical installations appeared that steel became used in a standard manner. The same went for reinforced concrete. Though concrete had existed in the 1820s, formal use of patented steel-backed cement did not pick up until the 1880s (Smil, *Creating the Twentieth Century* 170–71). It took aesthetically pleasing projects as well as grandiose ones to bring about a change of attitude. The Swiss Robert Maillart (1872–1940) designed and built some 40 concrete bridges in Switzerland, all visually appealing. A colleague of his in Paris, Auguste Perret (1874–1954), designed delicate facades all made of reinforced concrete rather than cut stone. Finally, in 1889, Gustave Eiffel (1832–1923) completed the tower that bears his name for the Paris Universal Exposition held that year. The goal was to show the possibilities of steel, even though bridges had already been assembled using the material (Eiffel also designed the internal structure of the Statue of Liberty).

As buildings grew in complexity, new features came to be added, some of which first appeared as novelties rather than necessities.

MOVING SIDEWALKS, ELEVATORS, AND ESCALATORS

The turn of the century was also welcomed with a universal exposition. In Paris, the various exhibits, though intended to reflect in part the eventful nineteenth century, in many ways focused on the acceptance of new technology into everyday life. Simply put, inventions of the nineteenth century, such as the railroad, the telegraph, and electricity-powered contraptions, were now accepted, even though the dynamics associated with such things were still developing. In fact, many of the exhibits included explicit applications for electricity, such as elevators, escalators, and moving sidewalks.

The 1900 Paris Exposition had witnessed improved elevators installed in the Eiffel tower (built in 1889). But what drew the attention of visitors was the presence of escalators. Initially patented in the United States in the 1890s and referred to as elevators (because they, too, ascended), these numbered 31 at the exposition. One in particular, installed at the exhibit space devoted to textiles and clothing, suggested strongly the idea of shopping while riding (Goetz 53). Unlike the elevator, which was often enclosed, an escalator allowed potential shoppers to see the goods as they rode upstairs. It would soon become a special feature in major department stores, where shoppers paid for the luxury of using the escalator.

A cousin of the escalator, the moving sidewalk also made a substantial impression at the exposition. A ride some 2.5 miles long suggested a leisurely future, and some 160,000 people paid to appreciate its features daily. Unlike the escalator, however, it found limited use, though later in the century, it was adopted in some public buildings, especially airports.

While it is almost a truism to say that the elevator informed the development of the tall building, Europe could claim few such tall contraptions. Prior to the post-1945 boom in downtown business areas, most buildings had elevators service 6 to 10 stories, and many residential buildings, especially in working-class areas, lacked elevators altogether.

In Germany, a particular contraption with multiple moving cars became a standard feature of office and government buildings. Department stores, however, came to view elevators as added elements of prestige to attract a rich clientele. French writer Emile Zola in his novel *Ladies' Delight* (inspired from the Bon Marché department store still in existence) depicted the efforts of the manager at wooing a richer clientele by adding a new elevator to serve one of the fashion departments. In reality, some managers went so far as to post a door holder on each floor to suggest the lap of luxury when using the contraption (Treischler 142). Hotels broke up floors to add elevators as a means of further attracting clients willing to pay the highest price for suites, but also to show up competition. In Switzerland, the Hôtel Beau-Rivage, made famous because Empress Elisabeth of Austria-Hungary died there after being stabbed across the street, gained further fame for introducing the first elevator in Switzerland. It was rapidly followed by other hotels. In fact, hotels had often had trouble filling upper floors because of the walk up the stairs this required. With the elevator, they could now charge more for rooms with a view. This was all the more possible because roads now brought far more travelers over improved roads and rail.

The design of city housing changed radically after World War I and followed prewar trends that were accelerated through the conflict. The most obvious case was that of the Bauhaus, which developed and thrived in Germany's Weimar Republic (1919–1933). Architect Walter Gropius pioneered the architectural school and employed simplified cubic masses, assembled asymmetrically and utterly unadorned. It rejected the traditional imperial prewar gilded style and argued that function should drive style. The Bauhaus style of architecture would proceed from certain assumptions: First, the new architecture was to be created for the workers, in a radical departure from urban dwellings that had typically been handed down to workers desperate for shelter. In so doing, this architecture sought to reject the dominance of middle-class ideals in everyday life, partly in reaction to the notion that it was those ideals that had contributed to the horrors of the Industrial Revolution. Finally, in a move similar to what the Enlightenment had done two centuries earlier by seeking out the Greco-Roman legacy to make its case for individual rights, the new architecture would return to the original Classical principles of Western architecture. Architecture as ideology came to emphasize modern materials with no decoration. Buildings soon became theories constructed in the form of concrete, steel, wood, stucco, and glass. They had to have a flat roof and a sheer façade, with neither cornices nor eaves. Any color was considered

The original Bauhaus building in Dessau, Germany, inspired thousands of others in European and American cities with its cool, clean style. The emphasis on glass and open spaces alongside barren surfaces and plain tones makes the Bauhaus style the epitome of functionalism. Source: Astra, Geneva, Switzerland, reproduced by permission.

eccentric and unnecessary, which is why white, gray, beige, or black came to dominate.

By 1924, mass housing was the great social issue of Weimar Germany; by 1932, no other country had built more housing for its workers. Most of the buildings for workers were built with tax money. As most of the architects adhered to the principles of the Bauhaus, the result was a classical form of rational social housing with open floor plans, white walls, no drapes, and functional furniture. Although the workers mostly did not like these new dwellings, in matters of taste, the architect acted as the workers' benefactor.

Team design was stressed, and architecture and art were fused in the factory murals of Paul Klee and Wassily Kandinsky. The political right wing denounced the new style as cultural bolshevism, and in 1933 Hitler closed the Bauhaus and drove Gropius into exile. In 1937, the stars of the Bauhaus emigrated to the United States, where they were welcomed with open arms. Gropius was made head of the School of Architecture at Harvard. Moholy-Nagy opened the New Bauhaus, which evolved into the Chicago Institute of Design, and Mies van der Rohe, who had become the head of the Bauhaus in 1930, was installed as dean of architecture at the Armour Institute in Chicago. Among modernist architects remaining in Europe, the Swiss Le Corbusier would have a lasting impact after World War II.

THE ROLE OF THE ROAD

The need for effective transportation in nineteenth-century Europe was often in response to anarchical developments. For example, in Paris, France, urban redevelopment in the late nineteenth century in the form of redrawing boulevards had failed to produce the desired effect, while the mix of private companies exploiting short routes (horse-drawn buses, then steam and electric trolleys, along with railroad services to the countryside) meant a slowdown in efficient urban travel. On time for the 1900 Universal Exhibition, Paris opened line number 1 of its subway system,

Berlin. Vor dem Brandenhurger Tor m. Tiergarten

A Berlin street scene in the early 1920s. The automobile and the trolley have prompted a redrawing of urban topography, appropriating most of the open areas. The diminishing number of horses signals the end of an era. Source: Contemporary postcard, author's collection.

thereby following the example of London (1863), New York, Boston, and Budapest. But it was streets and roads themselves that first required rethinking and, most importantly, resocializing. As the horse disappeared, new rules and regulations had to be devised to welcome its replacement.

Obvious though it may seem, the expansion of the automobile's usage depended heavily on new road developments. At the turn of the century, the appearance of cars on small country roads wreaked havoc in the farmland but was also difficult for drivers, who relied on all kinds of newly designed hats, coats, and goggles to keep dust, mud, and projectiles swept up by their wheels from impairing them: Car drivers often looked like early airplane pilots. Drivers also faced the risk of slamming into trees planted too close to roadways.

The risks weren't limited to obstacles but included hostile pedestrians and even authorities. In Switzerland, cantons Schwyz and Glaris banned driving on Sundays, while canton Graubünden forbade all private cars until 1927.

The country roads were also difficult to police and to maintain. In France, the notorious *Bande à Bonnot* ("Bonnot's gang") made heavy use of cars as getaway vehicles on small roads near the French border to flee the gendarmes (often confined to bicycles and horses) after committing crimes. When it wasn't the risk of bandits that worried travelers, the lack of signage and the poor quality of the roads also played a role. Private investors put up some of the first signs and affixed advertisements to these signs. The lack of information was but a facet of the whole need to socialize people to the road and in the streets.

The problem overall was that the car was generally seen as having priority over all other vehicles and pedestrians. This state of affairs dated back to when automobiles were actually hard to stop, for fear the engine might quit. Eventually, their speed and power simply gave drivers a false sense of superiority, even at low speeds. To remedy this, the British government pursued for well over three decades a campaign calling for mutual responsibility between pedestrian and driver: to cross within the painted lines (the way the Beatles did on the cover of their *Abbeyville Road* album) was the best way to avoid accidents. Though this sounds obvious, limited studies of effectiveness, the need to cut costs, and a lack of understanding of how best to proceed in reaching the public meant that national campaigns had unclear results. Multiple solutions were tried, but the most standard ones have been kept and include flashing warning lights and white stripes on the ground.

It is also in the interwar years that the notion of rapid car transit or thruway flourished. Paralleling the development in the United States of the first turnpike (I-76 in Pennsylvania), Germany initiated a major highway project in the 1930s. Though the result paralleled the American endeavor, the intent was quite different. In Germany, Hitler was in power, and the highway system, though based on ideas developed in the preceding democratic Weimar Republic, was developed for propaganda and military reasons. The tracing of the highway paths in particular was done with an eye to ensure an easy movement of military divisions within the land in preparation for war. The presentation of such intent to the public, however, was that of an idyllic road that would swiftly bring deserving German workers to nature spots or to visit one another in the Nazi-structured community. The autobahn, as it is known internationally, remains famous for the high speeds allowed on it, though some restrictions do exist out of safety and environmental concerns (Zeller).

After World War I, facing the prospect of reconstruction, many European countries looked to the United States to draw ideas on how to best regulate a traffic flow that was expected to increase. Aside from Germany, the United States was the only nation to have constructed highways prior to World War II. The cost of the investment, however, seemed excessive, especially at a time when gas supplies remained rationed (until the early 1950s in some nations). Echoing an American study published at the International Road Congress in 1930, a French columnist argued that

> to cross some areas of Paris, the digging of either underground road tunnels or viaducts will become the only viable solution that could solve our traffic problems. Generally, though, American road practices make clear that investing in road building is essentially a productive undertaking. (Borneque 205)

Though highway building would eventually become standard fare in all European countries, the impact would be similar to that on American cities: traffic congestion, increased pollution levels, and the ever increasing need for more roads to more places. The difference, aside from scale, would be in the provision of alternatives to cars, through subsidized railways and mass transit (see chapter 3).

TUNNELING AROUND EUROPE

If roads made car travel easier, tunnels in the city and beyond broke physical boundaries. Though they were very much part of nineteenth-century civil engineering tradition, longer arteries are often associated with the twentieth century. For example in Switzerland, the Matterhorn tunnel inaugurated the century's tradition of great tunnels. Though projects had existed since 1886, cantons who would have been involved in the digging of a tunnel under Switzerland's highest mountain were hesitant to commit to such an endeavor. Eventually, approval was given and construction begun in 1898. The lives of miners involved in the piercing was notable on several levels.

At the technical level, the loss of life in the piercing of an earlier tunnel (the Gothard's) had made it imperative to devise new methods of work. Every 150 feet or so, temperature increases 1.8°F, and suffocation becomes another danger miners face. New high-pressure water pumps were invested in to ensure a bearable temperature where the rock was drilled (temperatures there could reach in excess of 158°F). Workers were also subjected to pressures reminiscent of the bends for scuba divers. The dusty air had to be evacuated using newly designed air pumps, and engineers had to contend with a daily risk of deformation of steel beams under the weight of rock caving in.

At the social level, such daily concerns help explain not only the worries but also the commitment workers and their families had to the completion of the Matterhorn tunnel: It wasn't just a national project; it was *their* project. The state of mind among the workers and in the adjoining regions when the piercing happened on February 24, 1905, reflects it best. Each round of digging usually yielded 5 feet of progress. On that morning, the finishing round yielded 8.2 feet so that the next one, according to engineers' estimates, might be the last:

> Under such circumstances it becomes hard to hold back the miners. Each team seeks the honor of the final piercing; work proceeds with great excitement. Deeper holes are drilled than the ones ordered. Stronger [explosive] charges are placed in some cases. . . . A little before 8:00 A.M. a phone call in from the tunnel stating the piercing was perfect. Before one could get extra details, we all heard in the narrow valley the strident whistles of steam locomotives announcing the great news to the whole region. Instantly, flags appeared on window sills and village squares. (quoted in Kovári and Fechtig 64)

The piercing of such links, first for railroads but soon for cars, too, changed modes of travel and opened new borders. Many more tunnels would follow these pioneering endeavors, but few would cause environmental concerns until the late twentieth century. The most famous ones included the Mont Blanc tunnel, which links France to Italy under Europe's highest mountain, and the tunnel under the Channel ("Chunnel").

The Chunnel is arguably the longest civil engineering project in history, considering the earliest projects date to the eighteenth century. A first attempt, begun in 1875, ended in 1882 for military reasons: A tunnel might provide an ideal commercial venue but also offer quick passage to troops in either direction. A century later, another attempt was also cancelled a year into its undertaking, primarily for economic reasons this time: Europe had just suffered the first of two oil shocks in the 1970s. These had sent the economies of all industrialized nations into a spin. The third full-scale project would prove a charm. Launched in 1987 by Prime Minister Thatcher and President Mitterrand, the Chunnel saw its junction completed on December 1, 1990. The first train service began in 1994.

POLLUTION

What these workers experienced soon became identified as two separate phenomena: industrialization and urbanization, whose effects often crisscrossed by creating pollution, noise, ground vibrations, and poor sanitary conditions. Most early industrialized workshops were located in living areas, even when they moved to bigger spaces on the outskirts of cities. Foul odors and smog followed, naturally, so that laws to deal with such troubles appeared to try and quell the onslaught of industrial fallout. More was yet to come.

The late nineteenth century witnessed the second Industrial Revolution, whereby progress associated with scientific discovery, especially in the realm of chemistry, led to the creation of new generations of factories and a new wave of associated concerns. This time, it was not the working class alone that suffered from breathing fouled air. The notorious smog associated with British cities found its equivalent in European continental towns, and new machinery requiring the use of fuel increased urban fears of pollution. The turn of the twentieth century witnessed a shifting attitude, whereby fresh air and drinking water, seen as welcome luxuries, became a natural expectation. This was reflected in industrial circles, whereby factory owners, though still intent on observing the bottom line, accepted the notion that public health had to also be considered when establishing a techno-industrial project. This does not mean that Europeans "thought green" but that early legislation did begin to appear to deal with the issues as they became more clearly identified.

A French law in 1810 proposed checking the conditions of any establishment that released foul smells, and the first British Smoke Prohibition Act appeared in 1821 to try and deal with the smog. Other countries followed, but the issue of enforcement remained unsolved and regional and urban authorities often took it upon themselves to institute inspections of industrial buildings. In France, though the practice dates back to the nineteenth century, it would not receive official sanction until 1917.

Enforcing these laws, however, remained difficult. Public authorities trying to determine whether the burning of a specific acid could cause atmospheric trouble around the factory relied on scientists, but the latter often based their conclusions on the material selected and provided by the industry itself.

In the interwar years, a more proactive stance involved attempts at making industrial centers more functional and assigning them what amounts to makeshift solutions. If a factory's smoke blew onto a school, the smoke stack's shape would be changed, but not what it blew. More effective, however, was the introduction of industrial zoning laws that called for particularly polluting industries to move to open areas away from human habitation. But this could be applied only to new construction. In France, for example, where a 1932 law banned industrial smoke, inhabitants of

Fig. 4. Le Creusôt in 1782.

Fig. 5. Le Creusôt in 1851.

The expansion of the industrial complex of Le Creusôt, France, over a 60-year period shows the impact of industrialization on the environment. Such growth was encouraged, and its consequences rarely checked. Source: *Engineering*, January 1898, p. 5.

several cities, notably Nantes, called for inspectors to come and shut down foul-smelling factories. The inspectors were unable to even investigate because scientific evidence was not obtainable objectively. What this example demonstrates, however, is the disconnect between legislators and their constituents who felt that the state of science and technology was sufficiently advanced to enforce the law, and field inspectors who discovered their hands were tied (Massard-Guilbaud).

SELLING INDUSTRIALIZATION: ART, DESIGN, AND PROPAGANDA

Artists and designers were not insensitive to industrial arts. The nineteenth century witnessed several attempts at incorporating the machinery design into everyday modern life. Among pioneers in that realm, the German *Deutscher Werkbund* of Munich had shown the possibility of a symbiosis between industrialist and designers. But the international situation, especially mounting tensions between France and Germany, caused such industrialist art to be labeled German, even though there was a French Society for the Encouragement of Art and Industry.

Among the movements that technology inspired, Futurism stands out in its embrace of the machine, especially the speed dimension associated with it. The leader of the movement was Filipo Marinetti (1876–1944), who argued that the expressions of industrialization as exemplified by the automobile and the airplane were true harbingers of the future. Not only that, but it would remove all traces of a slow and boring past: "It is for Italy that we now establish Futurism with this manifesto of overwhelming and burning violence, because we want to free this country from its fetid gangrene of professors, archeologists, antiquarians and rhetoricians." This excerpt from the *Manifesto* of Futurism appeared in the French newspaper *Le Figaro* on February 20, 1909. Though seemingly welcoming a new technical age, the element of violence was very clear. Marinetti and his followers felt that war would act as the great cleanser and establish a clear, male-dominated moral order.

The way Futurists expressed their movement varied with each realm. It was difficult, for example, to devise a new form of dynamic painting, but they ended up heavily inspired by cubism, which was beginning to make headway in the Paris avant-garde. In other realms, there were attempts at Futurist poetry and even music. In the latter case, Luigi Russolo (1885–1947) tried "bruitism" (a concert of noise machines) to illustrate his *The Arts of Noise* manifesto. Finally, architects also sought to devise futurist buildings. None really came to fruition—the main protagonist, Antonio Sant'Elia (1888–1916), was killed in World War I—but the completion of the Fiat factory works in Turin in 1921 exemplified the notion. It included heavy use of cement and the completion of a flat roof top on which a car runway was built for testing vehicles assembled below.

Overall, Futurism would be redirected because of World War I and the eventual advent of Fascism in Italy. First, mechanized warfare changed European attitudes toward science and technology in a manner that the machine came to inspire as much awe as it did admiration. Furthermore, the Great War spawned a new political movement, Fascism, which shared with Futurism the emphasis on a male-ordered society that rejected old movements. Though Marinetti would later dissociate himself from Fascist dictator Benito Mussolini, the latter made heavy use of the modernist notions of speed expressed in Futurism. The osmosis of ideology and art was reached in part in 1929, during the final phase of the movement, when the "Manifesto of Aeropittura" was published, which glorified flight and its associated experience. Futurism also influenced other movements, including Rayonism in Soviet Russia, Vorticism in the United Kingdom, and even some of the Dada movement.

The Industrial Revolution also marked a turning point in the design of objects for daily use. Beyond heavy machinery, a new material culture made of domestic objects flourished. The most obvious reflection of this new facet appeared in universal expositions, which were designed to display what countries could do best with their new mastery of industrial technology. One way for consumers to examine new products and imagine the future was to visit the many national and international exhibits held in major European cities.

Universal expositions themselves were derived from the tradition of the great fair, some of which traced their origins into the early modern period, when preindustrial guilds staged gatherings. Some science and art groups also chose group exhibits as a way to promote their interest, but also to show the flag.

In the German states prior to their 1871 unification, the customs union that had appeared became a means to promote early industrial goods and suggest a German self-awareness. In Berlin, a grand exhibit took place in 1844 that allowed over 3,000 German exhibitors (most of them from Prussia) to show their production and advertise its quality. Subsequent exhibits (they were supposed to happen every five years) failed on political grounds and also because of the enormous publicity another exhibit drew in 1851 in England.

The Great Exhibition in London remains a symbol of the purpose of such shows. Intended to reconcile visitors to both the importance of the industrial age and to instill pride in the United Kingdom's domestic achievements, the Great Exhibition also offered a lasting impression through its architectural masterpiece, the Crystal Palace.

Of the 245 designs submitted for the exhibition hall, none really satisfied the organizing committee chaired by Prince Albert, Queen Victoria's husband. An outsider, a horticulturist named Joseph Paxton, was eventually chosen. Experienced in the design of glass houses, he offered a similar design for the great exhibition hall, and the grandiose building's size was only limited by manufacturing constraints of the time. This meant that the largest lateral separations between vertical beams would be no more than four feet, the widest size available for glass panes, for example. When completed (in 17 weeks using prefabricated parts), the building had some 4,000 tons worth of glass, about one-third of British annual production. Moved after the exhibition ended, the building stood in South London until its destruction by fire in 1936.

The purpose of the exhibit was to provide an osmosis of the arts and industry. In so doing, the 1851 exhibition set the tone for similar events in continental Europe and in the United States. Such exhibits would eventually prompt a new form of tourism. Such industrial exhibits, by combining with cultural achievements, could, it was thought, offer the ultimate form of peaceful triumph. Sixteen years after London, the 1867 Paris Universal Exhibition offered just such a reflection by combining displays of luxury goods alongside steel and other heavy industrial productions. Paris in 1867, and its later incarnations in 1878, 1885 (when the Eiffel tower was erected), and 1900, was also about entertaining. Technology as embodied in a "clean" setting of an exhibit was to be seen, appreciated, and even enjoyed as a symbol of a country's achievement. Such exhibits dovetailed most notably with the ones celebrating the centennial of the United States and developed a popular culture devoted to the festive element associated with such grandiose displays. Board games, medals, and later postcards and booklets all became part of the materials that entered the home, often on informational and educational grounds, yet just as often as simple souvenirs of a fun day at the fair.

Aside from universal expositions, new gatherings devoted to industrial and decorative arts featured new designs in home living, including associated machinery. Such exhibits, notably the Parisian International Exposition of Decorative and Modern Industrial Arts, also hinted at tensions in the art world over what was acceptable in terms of modernity and technology.

Other cultural reflections of Europeans' awareness of technology included new trends in architecture and the arts. At the International Exhibition of Modern Design in Paris in 1925 visitors were very impressed by the designs of Le Corbusier and Mies van der Rohe, who were heavily influenced by the Bauhaus ideas. In addition, other designers' works, while seeking to emphasize a national identity in each building (countries sponsored specially designed pavilions), all joined in the experimentation of new functional forms. Perhaps responding to this new trend, that same year, Walter Gropius published *Internationale Architektur*. This seminal volume identified the process by which a new style inspired itself from the creation and operation of machinery. In the European context, this became known as the International style.

Combinations of all three trends we have considered also existed, especially after 1945. Inspired by developments in Sweden, after World War II architects and commentators in Britain promoted a modernism that also drew upon older British ideas and traditions. In their combinations of modern architectural lines and structures with more traditional British features and materials, the South Bank Exhibition and the Lansbury Estate embodied and promoted this way of thinking. While the ideas may not have been brand new in 1951, the festival was a vehicle for both trying them out in the real world and introducing them to the public. London was the center of the Festival of Britain. The festival was a platform for new British art, science, and architecture on a grand scale. A postwar marvel, it was forecast to be a significant tourist attraction and boost to Britain's tourist economy.

URBAN REBUILDING AFTER WORLD WAR II

Though many smaller cities were bombed or destroyed in World War I, it was the aerial bombings and street fighting of World War II that took the greatest toll. Strategic bombing, though ostensibly aiming at war industries, often ended up destroying civilian urban dwellings. Consequently, the end of hostilities saw a dramatic need for housing displaced civilians and returning soldiers and gave new impetus to implement urban programs that would take into account population growth but also balance quality of life and rebuilding efficiency. Cities suffering substantial damage faced housing shortages and rushed to build dwellings as fast as possible. Other places looked to the United States for guidance, especially in road planning.

What they saw was too many expensive wish lists that the small volume of traffic did not warrant. In 1954, for example, whereas the United States had one car per 3 inhabitants, Sweden only had one car for every 13 (Lundin 323). In the latter nation, a special traffic and parking commission was established to try and account for future car travel and ended up mandating certain building codes that allowed for future traffic growth. In the case of Sweden, the adaptation of U.S. standards (with some adjustments) was deemed a success. One reason for this included a scientific approach to traffic planning that borrowed easily from published American models. For the average Swede, by the 1960s, a car society was in full swing and included the affordability of smaller road vehicles that could use all the newly constructed roads and parking spaces (Lundin 326–27).

As far as housing went, the legacy of the interwar years was substantial. In Germany, for example, Bauhaus-style modernist architecture returned, and in Europe, in general, several urban areas were modeled on the designs of Le Corbusier. The European city, however, did not follow the American model in shifting to a business center with a suburban residential ideal. Instead, in a move that echoed the initial urban migration of the nineteenth century, new city blocks were built outside of major urban centers and linked to them through public transportation. These "dormitory cities" as they were known became one of the themes of protest in the 1970s. Indeed, whereas the buildings themselves were generally of reasonably good quality and design, no thought had been given to what critics called the soul of the city. Thus, the buildings were often empty during the workday, and in evenings and on weekends, inhabitants found themselves isolated from family or leisure activities, from market to public shows and youth activities. The limited availability of shops combined with the colorlessness of buildings to create a depressed atmosphere. When framing this architecture in the context of poverty, immigration, and unemployment (some of these buildings were subsidized housing), it comes as little surprise that crime became a central concern, and social uprisings occurred on a regular basis. From Marseille's Cité Soleil, to Bonn, Germany's Tannenbusch sector (sometimes nicknamed Tannenbronx in reference to New York City's rough areas), European nations' social services have experienced substantial difficulties in dealing with the problem. Here, an architectural ideal that was to make urban life better actually failed, as reflected in popular songs that decry suburbia and the technology that brought it about. It would also feed the beginnings of an organized Green political movement.

CONCLUSION

The advent of the modern city in European history became a central feature of the Industrial Revolution. It thus defined Europeans' relationship to technology and to each other as new machinery appeared to deal with the concentration of the population. Some of these machines acquired a dynamic of their own (see chapters 3 and 4). Urban culture, too, developed a dynamic of its own as reflected in art and architectural movements. At the level of mass culture, new concerns also arose, as youth came to experience transnational influences rather than typically regional ones. This was also reflected in intergenerational conflicts, some of which were crystallized in the rise of ecological consciousness. For the grand majority of urban dwellers, however, urban culture also signified a dreariness that was not always reversible. Whereas downtown areas in most European cities remain a desirable (and expensive) place to live, the low rent dwellings on the outskirts of cities have brought about socioeconomic problems similar to those American cities encounter. Thus, urban culture continues to serve as an ambivalent feature of European cities, but the dynamism associated with agglomerations remains strong.

3

THE RAIL IN ALL
ITS EXPRESSIONS

ENTER THE IRON HORSE

James Watt had perfected the steam engine in 1768, but it was not until 1804 that George Stephenson was able to test it on rails. In 1829, he introduced the "Rocket" steam engine, able to function for lengths of time and draw carriages without suffering failure. Yet, the new technological system the British inaugurated called for a multitude of parallel developments. Initially considered as a replacement for animal-drawn wagons in mines, the transformation of the locomotive into a full-fledged railroad called for the development of new metallurgical solutions. New rails, but also ballasts to stabilize them, as well as better-designed pressure chambers to avert explosions, not to mention costs, all proved to be challenges. Yet the railroad became the first means of technological mass transport in the nineteenth century, and it spawned variations within cities.

Railways fascinated governments and masses alike. While they could successfully integrate economic structures in the course of the Industrial Revolution, they also offered a strategic advantage for troop movement. But railroads could not have been built without a unique combination of government involvement and private enterprise. Special investment banks were set up, and private industry in the form of engineering groups and iron, steel, and coal companies all came together to help build the railroad. In so doing, they also relied on it for their own expansion. In German states in the 1850s, over a quarter of all pig iron was used to build railroads (Trebilcock 43).

THE RAILROAD

Bringing the railroad into Europeans' lives was no small affair. Not only were infrastructure problems to be resolved (see the tunneling summary in chapter 2), but also the matter of *who* paid for it all was central. The nineteenth century saw the application of liberalism, an ideology the European middle class favored and that emphasized private enterprise. The problem of costs, however, was such that it became difficult to turn a profit on such large investments. Attempts to have the state take over the building of rail tracks for both planning and practical reasons were soundly defeated. In Switzerland, a leading parliamentarian put forward a now familiar refrain that "private companies manage better and more efficiently than the state does" (Kovári and Fechtig 15). The motion to nationalize Swiss rails would be soundly defeated in the face of his argument, and it would be over fifty years before a Swiss federal railroad came about. In the meantime, a new social development came about with the advent of railroads.

Train stations reflected the second Industrial Revolution but also the second generation of railroads. Built around the turn of the century, such stations were cathedrals of transportation. Many were built on the site of earlier stations, demolished due to small size or burned down accidentally. The purpose of such grandiose building was not only to celebrate mobility, but also to announce the importance of the city to the arriving traveler.

For example, in Switzerland, Basel's Bad station designed by architect Karl Moser and completed in 1913 was deemed the welcoming jewel for visitors from Germany. Main arteries led to it, and the station was less a fortress than a means of joining old and new. The train station in general had been considered a means to link downtown to suburbia, but in Europe, the movement went inward: Urban redesigns, notably in the great capital cities, led to train stations becoming intermediate links to downtown department stores built nearby.

Writers all celebrated the train station, calling it a "volcano of life" as it spat out hundreds of travelers, or even "cathedrals of a new humanity" (Schivelbusch 42). Railway atriums were places of encounter between travelers and their place of destination, and architects were careful to emphasize traits that might reflect an invented national character. Most stations designed with such ideas in mind appeared in the late nineteenth century, when states stepped in and nationalized private railroads for strategic, social, and financial reasons. Not only did train stations become national symbols, but the fascination their huge sizes exercised extended to toys, as manufacturers rushed into production tin models intended mostly for boys' train sets.

The nineteenth century had witnessed initial misgivings about using the railroad, but the expansion of train lines made it a central element of any

Train stations became important gathering centers in cities as well as small towns. Here, the inauguration of a narrow gauge regional station in Austria became the occasion of major festivities. Visitors and officials posed for the photographer, displacing the new building and the visiting locomotive as the centers of attention. Source: Stern & Hafferl, 1898, via http://de.wikipedia.org/wiki/Bild:Gurktalbahn_%28Eröffnung%29_retouched.jpg.

modern economy. In Belgium, a country newly established through the 1815 Congress of Vienna, the decision was also of a political nature. Most major waterways ended into the North Sea through the Netherlands. The possibility of linking industrial production to Belgian seaports became a central theme of the new railroad there. The United Kingdom was first sought out to help develop the new rail, but eventually France stepped in, and then finally a Belgian company was set up.

New steel and cement technologies allowed the planning and laying of new bridges across valleys, thus cutting travel time further. The notion of international travel and associated lengthened trips introduced several new features in everyday life. These included the use of the 24-hour clock on the continent as a means to differentiate A.M. and P.M. (and to acknowledge the time zone concept) and also the installation of onboard toilets.

Much of the impact of railroads was seen in the nineteenth century and involved a substantial cultural dimension. To governments, the locomotive was not just a strategic advantage militarily, but an expression of better control. In Sweden, laying the tracks in the northern areas was a way to link the metropolis of Stockholm to the reindeer farmlands, deemed in need of enlightenment (Hård and Jamison, *Hubris and Hybrids* 178). The

Locomotives often became the symbol of industrial power, and artists, writers, as well as the general public expressed fascination and fear when speaking of these. Here, an advertising postcard mentions the builder, Schneider, a major iron and steel producer. Source: Author's collection.

same went for the British as they laid their railroad in India. As rail travel gained acceptance, it became part of the themes writers might choose to immortalize. In Italy, Carlo Lorenzini published his experiences of a train trip in 1856. The author would become famous under the pen name of Collodi, writing the tale of *Pinocchio*. Charles Dickens described the impact of railroad building in *Dombey and Son,* while across the Channel, Emile Zola took on life among railroad workers in *La bête humaine,* while Robert Louis Stevenson's "From a Railway Carriage" passed into children's rhymes. In painting, too, the railway gained symbolic importance. In the arts, William Turner's *Rain, Steam, Speed* inaugurated the fascination many impressionist artists would record on their canvas.

Traveling also involved new advantages. Unlike horse-drawn coaches, railcars allowed one to move about from one compartment to another, as was typical in the British system. In Europe, a hybrid approach was developed, whereby some cars were open units, in the manner of ferryboats, while others retained the semiprivate compartment. The latter allowed night sleepers, especially in first and second class. The third class had none. As comfort levels increased (and so did cost differences between first and second), the third class continued to function (partly because of the impact of the world wars) but eventually disappeared. In 1956, all European railroads had cancelled the third-class option. Other amenities included the dining car. Once introduced, the latter made some line stops obsolete, much to the dismay of café owners and the delight of travelers.

The train world of course included train drivers who developed a specific identity that kept children in awe but was also shrouded in mystery. One train mechanic told a journalist in 1924 that locomotives were "like horses; there are good ones, bad ones, and OK ones . . . The mechanic, though, is like a horse whisperer; he has a lot to do with how well the machine runs" (Borgé and Viasnoff 31). Training to drive a locomotive involved as much work on the line as an apprentice as it did in the classroom. Practices also varied from one country to another. To keep a steam locomotive's pipes from freezing in the winter, French train drivers used either steam or cheap coke, saving the higher grade for propulsion. In Germany on the other hand, wood was burned. Even the type of high-grade coke varied. Coke that had not been filtered or cleaned burned off a lot of smoke and was kept for freight trains, while the best kind was reserved for passenger trains. A whole ritual preceded the preparation of trains for departure, similar to that found on American steam trains. Because the length of trains in Europe was on average shorter than American ones, trains had to draw less. But preparing any locomotive still took anywhere from one to three hours to get the steam pressure in order. The complexity of the operation was simplified somewhat with the arrival of diesel- and electric-powered locomotives, yet the time spent preparing for a train run (and ending it) remained an issue, as train drivers and mechanics ended up often working more than 10 hours in a row, which went against the law in many countries.

As for the rest of the train crew, like in the United States, many were assigned the same routes and became familiar to many regular travelers. A similar situation involved station managers who often lived on the premises with their family. Other nonmobile personnel, however, remained unknown to passengers: the rail attendants and crossing guards. The first were essential to the proper running of the railroad, as they maintained rails in and near train stations, repaired traffic signals, and offered extra means of visually checking that paths were clear. Their living conditions were generally very bad and only improved mildly when train companies built shacks for them to rest in bad weather. Though predominantly male, some women were hired, as their primary function was to clear rails from debris and the excrement that had dropped out from the train's toilets. Though the profession eventually shifted, it was an important feature of rail travel until midcentury.

Crossing guards became part of the European landscape as a means of ensuring proper crossing safety. Trains crashing into pedestrians and road vehicles was a common occurrence and fed public angst about train travel. Even though rail safety had substantially improved since the nineteenth century, gate crossing remained an issue. The guard was expected to step out of his or her house, lower the gates (usually with a hydraulic crank), and raise them as soon as the train had passed. In fact, well into the 1970s, some European regions still had crossing guards, and human error (such as not hearing the remote alarm that a train was approaching) played a role in several train collisions with road vehicles.

To ensure that the public would not react harshly to the fear of accidents, railroads set about improving travel safety. By 1935, radio had been installed aboard trains. Through electrical signals, these could also circulate in bad weather. Tunnel safety, in the form of better airing, was also established at the turn of the century.

One reason people were willing to use the train besides speed and long-distance coverage was that it had become cheaper. The reduction in fares in the 1890s had further contributed to making rail lines an attractive option for travel (Borgé and Viasnoff 181). A new culture of train travel began to appear, one in which proper ways of traveling were even suggested. The following excerpt comes from a French article that advised travelers on what to do and to avoid when getting onboard a train:

> Before climbing in, choose as much as possible, regardless of which class you are in, a car in the middle of the train, and a compartment in the middle of the car. This way, you will feel fewer vibrations, experience less noise, and sit further way from the wheels the movement of which echoes in compartments. If, despite this you were still shaken during the trip, it would mean that the train's linkage has a defect. In other words, the shock absorbers of your car are not in perfect contact with those of the next one. At the first stop, ask the conductor to have the hooks rebolted. He won't decline the request. Aboard the train, the best position, if your stomach allows it, is facing rear, sitting in a corner. This way, you can take on drafts and dust . . . Also, try to sit as close as possible to the alarm bell, so that you may use it swiftly when necessary. To climb into a car from a low sidewalk, place your right foot on the first door step, so that you may reach the car platform with the same foot. If you do the reverse and you are carrying luggage, you may lose your balance and your co-travelers might laugh at you. Such recommendation also applies to ladies who often become entangled in their skirts because instead of slightly bending their legs by raising them slightly, they raise them suddenly in front of them. Do not climb into a running train. This is forbidden. In train stations where the sidewalk is at platform level, do not step out early: you risk slipping or catching your legs under the wheels, unfortunately a common accident. You risk also slamming into a bridge support, a column or a traffic signal. Wisdom and train rules recommend one not get off a train until it has come to a complete stop. For those who would wish to go faster, we recommend stepping down forward, not backward, and to raise their foot on its tip slightly instead of laying it flat on the ground. Tip the body slightly backward and let go of the handrail. For women, we recommend a similar stance; only they should raise their skirts

slightly so these won't catch the first door step, which would cause a difficult and painful fall. One a train arrives, and since the breaks are usually set suddenly, hold on tight when standing, for you might fall on your neighbor and hurt him or her. Do not travel with a cane or an umbrella between your legs, but lay it sideways right or left. In case of collision you would hurt your hands and head. . . . There is also a proper way of stowing your luggage in the overhead nets. If you lay it poorly, you risk getting hit on the head or knocking out your neighbor. . . . Spread this advice far and wide; it will help avoid travel incidents which are caused by the traveler's inexperience. (*Lectures moderne* 1901; quoted in Borgé and Viasnoff 185–87)

The need to socialize consumers to the use of trains also saw the introduction of new rules pertaining to health. What one historian called "the pathology of railroad travel" (Schivelbusch) involved a whole series of ailments and fears. The first was puzzlement at the sensation of speed. The radical increase in sustained velocity affected visual perception. Maintaining speeds exceeding 20 mph, a locomotive crossing a field of flowers allowed its passengers to see only "specks of colors," as writer Victor Hugo put it. The medical profession expressed concern that such movement might hurt the heart or even the brain of travelers, and it was not uncommon to read of such health warnings in popular magazines.

Other concerns pertained to the injuries one could experience when falling on the tracks or standing too close to a passing train. A British official had even died during the inauguration of a railroad link, and the gory descriptions of dismembered bodies in early accidents made the fodder of early tabloids. The proverbial "train wreck one cannot look away from" derived its cliché from the true morbid fascination a technology gone awry exercised on survivors and witnesses. Consequently, all kinds of warnings appeared in relation to safe travel.

The famous "do not lean out the window" that appeared in several languages in European trains was but one of them. Another, "do not spit," though comical, had medical grounding, for up to the midcentury, the ravages of tuberculosis meant that exposure to the illness was common. People who were affected traveled to take in better air, and the disease, also known as consumption, forced them to expectorate infectious phlegm. Doctors recommended travel in specially configured coaches that could be easily cleaned and disinfected. Independently of the illness, the evolution of the railroad into a mass phenomenon required further socialization of travelers. A railroad car, though public, could not be considered a street. Ticket controllers were thus empowered to fine such offenses (Borgé and Viasnoff 191).

Other travel rules went unwritten, yet involved certain social rituals. How much did one acknowledge one's neighbors, and how many odd attitudes should one allow? Some abhorred men who took two seats, who talked politics, who read by candle light (or in later years with a flash light), or who removed their shoes. Finally, the snorers got the worst reputation. Though such a litany exists everywhere nowadays, the first lists of such unpleasantness appeared before World War I!

The most important issue when one entered a carriage was perhaps the trivial detail that concerned "ownership" of the window. Was the one who sat next to it empowered to open it and close it at will? Travelers in compartments were known to have gotten into shouting matches. The window-seat occupant requesting permission, on the other hand, might still be denied. The class in which this happened also had an impact, just as the nationality of the train and its occupants. A peer pressure of sorts also existed, whereby those who boarded at later stops were often condemned to put up with whatever position the window was in. To resolve it logically, etiquette experts argued, "the window belongs to everybody, be diplomatic; it is the only way to travel with pleasure" (Borgé and Viasnoff 185). This state of affairs continued well into the 1980s, when the introduction of high-speed trains and associated air conditioning sealed windows for good.

Of course, one way to avoid such arguments involved paying for a first-class ticket and seeking out a private compartment. By the interwar years, travelers who could afford higher prices came to expect more and more comfort, and railroad companies competing with each other upped the ante as much as possible. The Great Western railway in Great Britain set up bar compartments in 1934, for example, but in so doing, its managers were emulating previously introduced luxuries, such as those found on the Orient Express.

An important development of the use of railroads in Europe involved leisure travel. As tourism grew to mass proportions, the possibility of traveling around one's country or even abroad meant that only costs and vacation time limited the average traveler. Though middle-class excursions and travel were common prior to World War I, the social impact of the Great War changed travel habits.

When mobilization was declared in the European capitals, trains to the borders immediately dispatched standing reservists. Many a train bore the graffiti "A Berlin" or "nach Paris!" depending on the convoy's nationality and direction. The train stations on such occasions became the last point of contact between civilians and soldiers. Big departure points such as Paris's East Station still bear witness to the event through a series of commemorative plaques. In that case, the ritual would be repeated a quarter century later at the start of World War II. But village stations, too, became the site of separation, and the alternance of solemnity, sorrow, and encouragement of the crowds assembled to see off the men, punctuated by engine sounds and whistles, became part of a European landscape of memory specific to that continent. The same could be said for the return of the soldiers who survived. As Erich Maria Remarque describes in *The Road Back,* the station was the first expression of promise and delusion associated with the city, itself described as a woman full of promises and worries. Soldiers who had survived together several years in the trenches now separated and tried to make the best of their return to civilian life. Some were unable to adjust and, like in reality, returned to the battlefield, as tourists.

The first expression of a new kind of train travel came in the form of battlefield tourism. Widows, but also veterans eager to honor their fallen comrades, began seeking out means to travel to the trench areas, and rail companies eventually accommodated such requests. In parallel, however, redeveloping the railroad as an economic imperative became the top priority for all governments.

The train as daily routine should not detract, however, from the fact that glamour remained associated with that mode of transportation, in particular in the form of luxury lines. Such was the case of the Orient Express, worth citing because of the impression it made on the European imagination. It wasn't just Agatha Christie's Hercule Poirot who dealt with *Murder on the Orient Express,* but the very special level of comfort it offered, which affected a shift toward more comfort on regular trains.

First inaugurated in 1883, the train was the brainchild of engineer George Nagelmackers, who was inspired by the American Pullman cars and adapted the comfort levels to suit European tastes. Within three years, the train, which used to end in Romania, traveled all the way to Istanbul.

The prestige associated with the lines was commensurate with that of ocean liners and drew a peculiar fauna of diplomats, spies, and socialites. King Ferdinand of Bulgaria gained notoriety by stopping the train on his territory. He loved its features and, claiming the machine was in his home, boarded it in white overalls (quickly darkened by the smoke), stood with the drivers, and proceeded to test the machine: high-speed runs in sharp curves, sudden application of the brakes, and inspection of the compartments all would be anecdotal whim of a bored monarch, were it not for the fact that he did them many times! Complaint after complaint from crew and passengers piled up on Nagelmacker's desk until a compromise was finally reached, allowing the king to have his royal coach attached to the train on Bulgarian territory in exchange for letting the conductors do their job (des Cars 47).

By the 1930s, other cities were insisting on and hoping for an Orient Express line. Several such links were set up, but none gained the prestige of the original one. One likely reason was the impact of the train on literature. Graham Greene authored a novel by the same name. In it, chapter titles follow in the manner of train stops (Ostende, Cologne, Vienna, Subotiga, Constantinople) as the characters cross paths in the narrow corridors of the sleeping coaches. More than just the train itself, Greene actually showed a Europe seen by night from the Orient Express, thus increasing the feeling of mystery in the novel. These descriptions, albeit romanticized, fascinated readers by helping them imagine what a trip to the confines of southeastern Europe would be like.

Frenchmen Paul Morand and Joseph Kessel used the Orient Express more as a prop than an adventure. Morand in particular suggested "everything ends in literature, including the Orient Express" (Fedorovski 56). Other less well-known writers and poets also used it to gain inspiration for their writings. Part of the magic, at least up to World War I, involved the relative ease with which those who could afford it traveled: All one needed was a letter of introduction or even a business card, for border checks did not formally exist.

Movies also capitalized on the train's prestige. Some did so in an implied manner (Alfred Hitchcock's *Lady Vanishes*); others more openly (*From Russia with Love*, an early James Bond installment, and the Sidney Lumet-directed interpretation of Agatha Christie's Hercules Poirot investigation *Crime on the Orient Express*). Though initially an exclusively first-class service, second-class coaches appeared after World War II, and even third-class in the 1950s. The cold war as well as the competition from airlines shrank attendance dramatically. Finally, in 1977, the last regularly scheduled Orient Express train left Paris.

THE PRESENT DAY

Overall, the train in Europe had a similar impact to the American railroad, but the trend carries on today. Attempts at privatization of the railroad, as in Great Britain, occasioned substantial public commotion during the mandate of Prime Minister Margaret Thatcher. Most other nations chose to focus on maintaining national control over their railroads while encouraging competition with other means of transportation. Chartered trains with guided tours or visits to rock concerts, but also special pilgrim trips to Compostella in Spain or Lourdes in France, though in existence for decades, became advertised.

Finally, the introduction of high-speed rail starting in France in the 1980s with the TGV, and followed up in Germany with the ICE, also signaled serious competition for the airplane. For example, someone boarding a TGV in Geneva, Switzerland, could now expect to see the roofs of Paris 3.5 hours later. A similar trip aboard a budget airline now takes as much time when one accounts for traffic, airport lines, and luggage retrieval. Ironically, however, the design of high-speed train interiors was very much inspired from that of aircraft, though of late, the emphasis has been on new comforts designed to woo a variety of clienteles. The success of the high-speed trains, traveling on average well over 200 mph, should not mask that such success happened with heavy state subsidies, especially in the development of new infrastructure such as high-speed rails.

THE TROLLEY AND SUBWAY SAGA

The railroad was the technological catalyst of the modern city. We shall look at its acceptance and use in chapter 4. However, two of its derivatives should be discussed here: the trolley, essentially a narrow-gauge version of train transit, and its variation, which soon developed its own tradition and culture, the underground rail.

Though the trolley (aka tramway or light rail) is a nineteenth-century development that first came about in the United States, its evolution in twentieth-century Europe is remarkable. It represents a technology that displaced horses, introduced mass transit, and then was declared obsolete by midcentury and disappeared from many Western European areas, only to be reintroduced in the late twentieth century in response to economic growth and road overcrowding.

Prior to 1900, many Europeans living in cities were familiar with horse-drawn trolleys. Though slow, they were nonetheless fairly punctual and represented a logical extension of the railroad into the European city. The first introduction in May 1881 of an electric trolley in Europe occurred in Berlin, Germany. This event followed the prior introduction of similar contraptions in the United States. In terms of economics and technical advantage over horse-drawn carts, the electric trolley should have been introduced at once. Instead, the switchover took over 20 years because in most European cities people became upset with the prospect of overhead electrical wiring (McKay 86).

The reaction against electric tramways was extremely strong. City residents argued that the electric wiring would destroy the vistas of the city they lived in or would even present new dangers difficult to face. In the latter case, horrible though rare accidents of falling wires electrocuting passing pedestrians and even horses suggested the new mode of transportation was not yet mature. In fact, the electric trolley was a tested means of urban travel based on the American experience, but to gain acceptance, the judgment of engineers and the actual testing of trolleys in suburban areas was required. Such initiatives were generally the result of private companies vying for trolley contracts (McKay 111); though the negotiating took time, it eventually bore fruit after the turn of the century. By then, with the introduction by private companies of electrification, other operators felt it essential to modernize, too, and sought to withdraw horse trolleys. In the case of Zurich, it was not until 1907 that the city moved to acquire private companies to consolidate service and ensure a sound development of the service to account for regional growth. This decision, echoed in many European cities, was not just a matter of sound planning but of response to popular demand for better transportation.

After 1900, the trolley industry had matured substantially. Yet, it would have come to nothing were it not for the social impact of the tramway. As historian John McKay put it, trolleys were no longer an invention but a revolution. From a luxury, public transportation had become a necessity: On average, Europeans used trolleys four times as much daily by 1913 as they had horse-drawn carriages four decades earlier (McKay 193). Children hopped trolleys to school, while their mothers ran errands. Workers could afford to take the mass-transit system, something that had not been the case with horse-drawn tramways. Cities negotiated contracts with private companies to ensure proper coordination of schedules but also to maintain affordability. This last aspect of the introduction of tramways and trolleys is perhaps the most important.

The acceptance of the trolley also depended on urban administrations that subsidized trolley transportation for people of limited means. Though they dealt with private companies, administrators enforced rules of traffic, safety, and taxation. In the case of European trolley manufacturers who created many of the companies operating their wares, this was not deemed a hindrance, as the primary goal was to sell trolleys rather than operate them. Another reason for public intervention into private enterprise was the belief, based on nineteenth-century experience with industrialization, that municipal or even governmental oversight (in the manner of the railroad) was the best way to use new technology efficiently (McKay 245). The fact that the public in most European nations, or at least the educated public, was able to look beyond the costs associated with setting up such infrastructure also speaks to the attitude of many Europeans regarding technology: Though a matter of concern, its demonstration of service to the greater good could ensure acceptance. This was also reflected in the fact that in many cities the post office service, taking its cue from earlier practices in the United States, used specially constructed trolleys to service urban and outlying areas. Mailmen would then collect the presorted bags and complete delivery.

The trolley's heyday continued in the interwar years. By then, any horse-drawn transportation had entirely disappeared, and the combination of trolleys and bus lines in European cities (as well as subways in great cities) was helping manage the socioeconomic needs for mobility from inner cities to the outer industrial zones. The trolley in particular was credited with allowing people to move beyond traditional city limits. In so doing, it contributed to the development of a European suburbia while relieving some congestion in the city. However, the trolley also contributed to increasing property and land prices. It completed a revolution in mass transportation that had begun prior to World War I. Even the parallel introduction of buses and subways did little to displace the role of the trolley: In fact, the other modes of urban public transportation filled niches and paths already opened up. Another means of transportation, however, would signal the decline of the trolley.

The ever-growing presence of the automobile heralded the eventual end of the trolley era. As cars became more affordable and also gained acceptance in society, to travel locally in modern style came to mean owning one's own vehicle. Whereas in the United States Ford's legendary Model T achieved the central role of being "everyman's car," in Western Europe designs for affordable vehicles remained but prototypes whose further development was disrupted by World War II. The trolley had won a respite there.

With the rigorous oil rationing that accompanied World War II in Europe, street cars often became the last means of travel in cities and even in the nearby countryside. Viennese survivors of strategic bombings were often advised to bring out bodies for transport to the central cemetery. A specially modified trolley circulated on all the lines still in operation and loaded coffins sideways into the modified passenger space.

Following World War II, the destruction of much of Europe's economic infrastructure helped explain the increased reliance on public transportation but also the wish to capitalize on the tragic destruction of cities to rebuild better means of public transport: The trolley's days were numbered in some countries, while it continued to do well in others. Because trolley depots were set up throughout cities and in the outer areas, many cars survived the destruction and were put back into use once rails were cleared of rubble or reinstalled.

In German-speaking countries and in central Europe, trolleys continued to run. Removal costs, but also practical concerns, meant that most cities kept old trolley lines and even added some. The Communist-dominated regimes there also argued that such transportation was the most democratic means of travel. While partly true, it was also one of the cheapest, as the infrastructure was already in place for the trolley's use.

A comparison of the two Germanys is actually quite telling when seeking to understand the logic of public transportation under different political conditions. The Allies occupied Germany in 1945, but the beginnings of the cold war combined with currency reform in the western zone and the Berlin blockade among other factors led to the formation of the Western Federal Republic (FRG) and the Eastern Democratic Republic (GDR).

In both areas prior to 1949, urban experts came to view the trolley as the main means of travel in the city. In the western zone, factories received orders for new designs as a means of completely updating the transport grid and making it both attractive and efficient to users. In the eastern zone, where financial conditions were less propitious, a few new cars were produced, but most were rebuilt, and the designs remained closer to the kind used up to and during World War II. The only city where the trolley actually suffered a shutdown was West Berlin. Because the political situation no longer permitted its circulation eastward (the subway could still run, though some stations were sealed off), its activities were slowly reduced. The construction of the Berlin Wall accelerated its demise, and in 1967, the last operating western cars were withdrawn from use. Across the wall, however, new lines were put in, and because East Berlin, as the capital of the GDR, was to serve as the jewel of the state's urban efforts, newer trolleys were installed, too.

In Austria, where the city of Vienna had also been divided during its 10-year occupation, city authorities nonetheless proceeded to rebuild the damaged trolley lines. Beginning in 1955, new cars and rails were added, and nowadays, Vienna's trolley offers one of the most thorough coverage of any city in the world. This does not mean that the Viennese success was repeated elsewhere in Austria. While Graz, Linz, Innsbruck, and Gmunden all continued to offer trolley service, smaller cities such as Baden, Klagenfurt, and Sankt-Pölten eventually shut down service (the latter not until 1976) because economic conditions could not justify even a heavily subsidized service.

Though many European cities immediately set about planning a public transportation grid, the costs of building an infrastructure required that choices be made. When considering buses, subway lines, and trolleys, the latter were often deemed superfluous because of limited speed and how they impacted car traffic in congested arteries. Consequently, lines that had been destroyed by war were either abandoned or their frequencies reduced. By the late 1950s, France had done away with much of its trolley infrastructure. The city of Grenoble in the eastern part of the country did away with trolley service in 1968, when it hosted the Olympic winter games and wanted to show a modern face to the world. In the 1970s, city governments seeking to cut back their budget in the face of an increasing recession were told to remove trolleys from the rolls and concentrate on roads and parking.

In Geneva, Switzerland, trolley rails were either ripped out or cemented over in the late 1960s and replaced with electrical and gas-powered bus lines. A single trolley line was kept for practical reasons (it circled the city). Though maligned as "the line of the elderly," number 12 became part of the city's identity and culture. Because several sections of the line were in reserved areas where vehicles weren't allowed, the slow speed of the trolley (on average 15 miles an hour) was less of a hindrance, for traffic jams meant crossing the city by car could take even longer.

Some cities, notably in Austria and Germany, when faced with a choice between trolley and subway, chose a middle solution: the underground trolley. Only Hamburg abolished the trolley outright. The reasons for maintaining the lines varied from cost to political pressures to landscaping issues. The advantage of such a solution was to use existing cars without having to select a new gauge and order a whole new park. Also, tunneling in some areas would save travelers some time. Unfortunately, inexperience often affected the end result. In Vienna, for example, several tunnels were dug for the trolley lines. In one case, where reconstruction lasted three years, the rails entered and exited the tunnel at major road traffic junctions so that the time gained while underground was essentially lost when reemerging in open air.

By 1980, the trolley was generally viewed as a symbol of the past in Western Europe. Its strong presence in Vienna added an aura of aging to the Austrian capital, as it circled and crisscrossed the city in all directions. German cities in the throes of urban renewal argued about maintaining services and generally did so for reasons of costs and practicality. An 80-mile light-rail service between Duisburg and Krefeld, simply known as "the K," even had a dining car attached to it that allowed evening commuters to have early dinner. The dining service is gone, but the line still runs. Wherever the trolley survived, the public did enjoy using it, because it was considered safe and ran with the most regularity after subways (except when occasional traffic collisions interrupted a whole line). Its relative slowness actually encouraged street singers to use it to offer a tune or two and collect change from travelers (a practice often banned).

Nostalgia also dominated the symbol, not only for trolley fans, but also a public exposed to popular songs that made the postwar charts. Whereas the Beatles simply alluded to a trolley ride to buy a ring in "Ob-la-di," Belgian singer Jacques Brel constructed a whole love story around Brussels' "Tram 33," while in France Mireille Mathieu interpreted her own "trolley song." In the early 1980s in Germany, Reinhard Mey depicted the routine of trolley travel in "Das Lied der Strassenbahn," while in Switzerland, the blues group Beau Lac de Bâle reminisced about a flirt aboard

The trolley made a comeback in Europe beginning in the 1980s. However, it had never left some areas and had already been incorporated successfully into urban infrastructures, thus providing an excellent alternative to private vehicles. Here a trolley leaves the Rotterdam central station in the Netherlands. Extensive parking space is available for bikes, as commuters hop on rail transports to get to their places of work. Source: Author's collection.

Geneva's trolley 12. The trolley was thus immortalized to several generations, and it was far from dead.

Anecdotal though it appears, traffic congestions and extra pollution sowed the seeds of the trolley's return. The rise in environmental awareness throughout most of Western Europe in the late 1970s, combined with the double oil shocks of 1973 and 1979, prompted city administrations and national governments to consider cheaper and less polluting transportation alternatives. The jump in car ownership completed the resolve either to keep or to modernize existing trolley lines and introduce new ones.

Grenoble, which had ripped out its trolley rails in 1968 to welcome the winter Olympics as a modern city, put new rails in by 1988. Some twenty cities in France had followed suit by millennium's end. The point was to make downtown livable again with roads closed to cars and pedestrian zones installed everywhere.

In countries that had never ripped up their lines, partly for financial reasons, the return to light rail gave an added boost. In the Netherlands, coordination of trolley, train, and private transportation (especially bicycles) has proved an ideal solution to traffic congestion and eases the stress of getting to work.

UNDERGROUND TRANSPORTATION

The nineteenth century witnessed the beginnings of underground transportation, with cities such as Paris and London inaugurating subway systems that compared to those of New York City and Boston. Though the oldest system was that of Budapest (at the time part of the Austro-Hungarian empire, now capital of Hungary), the gestation process in the city of London is among the best documented. By the 1830s, the city had become so congested that the establishment of early railroad transit (above ground) required most end stations to be located outside urban areas, as added construction would have been prohibitive. Witnesses testified in the 1850s that the open-air rail solved little by way of traffic congestion and that it could take longer to cross London than to go from London to Brighton.

The suggestion to build an underground rail already existed in the 1830s, but it was not until 1853, on the urging of a private company, that the first line was approved. It took 10 years to build, but when it finally opened on January 10, 1863, it drew a crowd of 30,000. The carriages were often steam engine-drawn, but starting in the 1890s, electricity became the means of clean traction, through the use of a third rail that fed the engine. The resulting fumes were accepted as part of the experience, but in the early stages of operation, it was clear many passengers felt ill.

Among the social impact of the underground, one should consider its influence on the working class and the poor. Displacement of poor inhabitants from condemned buildings was tremendous but saw little to no protest. Initially operated through private companies, the London underground turned out to be far too expensive for many workers seeking to get to their jobs. However, several companies eventually lowered the return fare, which eased the budget of many blue-collar workers. This culminated in 1883 in the Cheap Trains Act legislation, which forced all railroad companies to offer special working-class fares (Freeman and Aldcroft 149).

Building/digging new lines varied considerably from city to city. Starting in 1900, for example, London saw a remarkable increase in the number of lines all the way up to World War II (during the war, its stations also served as underground bombing shelters). Excavation ceased for over three decades as the underground went above ground in the suburbs, but car traffic congestion in the 1960s and 1970s encouraged some new digging, notably to link nearby stations from various lines. Nowadays, the London lines reach over 700 miles and carry an average of 600 million passengers a year. The limited investments consented for upgrades, however, has made the service notorious, notably during the summer: No air conditioning was ever installed. Nonetheless, the so-called tube became a standard feature of London's identity. One reason, aside from convenience, was a substantial redesign in the 1930s. No longer were entrances dark stairways nor were waiting areas dank basements. Instead, added lighting and the covering of walls with glazed white brick sought to make the travel experience if not pleasant, in the very least bearable. The fashioning of standardized lettering, initially designed in 1916, became part of the tube's identity and is still used today.

Finally, to ensure ease-of-use, a new mapping system was introduced. Up until the 1930s, the London railroad lines, while marked with color, sought to follow a rough geographical outline to scale. This proved extremely unwieldy and confused travelers seeking to understand where they might connect from one line to another: The details were far too small to read. Harry Beck, a former employee of the London Underground company who had lost his job due to the economic recession, submitted in 1931 a design for a more schematized map. The diagram, which emphasized straight vertical, horizontal, and diagonal lines, removed the map scale by compressing lengths. Rejected at first, the design was accepted a year later on a trial basis. By 1933, the first run of 750,000 registered tremendous success with the public, and Beck's work, modified regularly, remained in use for the next three decades. Most public transportation maps nowadays are derived from this first synthesis (Garland 19–22, 60).

Not all cities avoided modernization. Glasgow, Scotland, for example, whose nineteenth-century subway was bought from a private company and electrified in the 1930s, underwent a thorough modernization in the 1970s. The claims that accompanied the refurbishment, namely on-time operations and the orange paint that covered all coaches, led locals to nickname the system Clockwork Orange (Bennett 39).

Major improvements also accompanied Budapest's underground after the war, but these were slow due to the need to allocate construction workers to housing projects to rebuild the war-torn city. Inhabitants and tourists made steady use of the system, but in 2002, numbers fell for the first time in the subway's existence as cars became more affordable and oil grew cheaper.

CONCLUSION

The railroad, as many historians have pointed out, was the first technical system to introduce Europeans to new conceptions of time and space. In so doing, it also introduced travelers to the notion that they could experience new destinations and thus new freedoms. The paradox, however, is that while bringing strangers to new lands, the railroad car also made strangers within it. New rituals limited the contact between travelers in a manner that had not existed previously. Such discipline was accompanied by a new attention to time. The clock had already changed work traditions in the late Middle Ages, but the train enforced these into rules. Missing a train meant missing work. Traveling thus served leisure but also imposed a servitude of sorts through its routine and time management. This is why personal ownership of some forms of transportation, namely a bicycle or an automobile, became so important in the twentieth century.

4

SELL YOUR HORSE AND BUY A CAR

MOBILITY AND THE ENGINE AGE

Whereas the railroad introduced Europe to rapid communication and changed the way entire nations worked, other means of transportation personalized the travel experience and affected smaller businesses' way of surviving. The nineteenth and twentieth centuries confirmed and deepened the notion of mobility in everyday life around the world. In Europe, the automobile introduced an evolution of social mores but also reaffirmed the value of public transportation, in sharp contrast to the United States. We begin with the vehicle that inspired newfound freedom as individuals learned to control not only their vehicle but where they went. Though the car would soon follow the bicycle, it took several decades before it became an affordable vehicle, in the manner of Ford's Model T in the United States.

THE LITTLE QUEEN (OR THE BIKE)

The invention of the bicycle can be traced back to contraptions that relied on foot thrust to propel two or four wheels in the early nineteenth century. Several variations in Germany, France, and Scotland appeared in succession. The first bikes to appear in Europe were the so-called high wheels of 1853 where the front, wider wheel was powered by foot pedals. Earlier projects and models had met with derisive comments, notably that these were but toys with no value. First made of wood, bicycle wheels soon received rubber covers and inflatable parts (the work of John Boyd

Dunlop), but it wasn't until 1891 that Michelin introduced the removable rubber tire. The bicycle eventually became one of the most used means of transportation. Tested in the 1870s in most European armies, by the turn of the century bicycle units served as rapid couriers.

The bicycle did not have the strength of a horse nor the autonomy of any farm beast. In fact, it represented a new kind of mobility at a time when sports had begun to reappear after centuries of nonpractice. The need for fresh air away from polluted cities, as well as that of proper exercise to ensure better physical fitness, overcame old Christian dogma that had condemned sports as a form of pagan vanity (a direct reference to the Ancient Greek Olympics). When Pierre de Coubertin and his committee staged the first modern Olympic games in Athens in 1896, track cycling was included.

Races became a central feature of bicycling as soon as the contraption was introduced. Though initially favored by the middle class alone (becauce bikes were fairly expensive), races eventually drew all walks of European life. Entertainment of this kind quickly became an important feature of modern sports but also of surrogate nationalism as spectators gathered to cheer riders.

The use of the bicycle fascinated the European public, but in urban spaces it did raise concern about proper public behavior. In Germany, for

The bicycle as a social outlet. A sports organization in Nuremberg, Germany, advertised its membership's fun with a collection of snapshots of bike tricks. Source: Postcard, author's collection.

example, cities required bike users to be registered and to have a numbered license displayed on the front of the bicycle. There was no such requirement for horse riders. The reason for this may have had to do with the far greater discretion the bicycle afforded its rider.

The opportunity to get away was the biggest draw the bike could offer. In middle-class circles, where the emulation of noble standards of behavior also restricted younger generations, the latter saw the bicycle as a chance to frolic in the countryside away from parental control. Boys in particular felt that this was an ideal way to show off and meet women who, likewise, might use bicycles to escape their families for a few hours. The German "Wandervogel," youth associations of high school and college-aged men and women, came to use the bicycle as a standard way of assembling and self-identifying: The two-wheeled contraption was as much a symbol of youth identity as their songs and rituals. Across the Channel, on the other hand, not everyone saw the bicycle as a positive tool. Reports of women being arrested for pedaling in the countryside resulted in charges of indecency, usually because the hapless rider wore pants.

Generally, though, the fascination with the bicycle in the news pertained to racing. The first recorded race took place in Paris in 1868, and several other countries, including the United States, followed suit. It was not until 1903, however, that the Tour de France began. Meanwhile, the popular love for the bicycle (nicknamed "little queen") prompted the creation of bicycling associations to review the calendar of events and provide advice to riders. At the first tour, only a third of all entrants finished. The following year saw multiple instances of cheating, while spectators were observed attacking some bikers or even dropping nails along the race roads. This latter instance remains a constant source of concern for organizers. The very nature of the race witnesses a ritual of campers, and passersby settle along the road to cheer or jeer their favorite biker. Though limited violent incidents do occur.

Like in any sport, rules evolved in response to events in the competition. For example, by World War I, riders could not hope for any kind of assistance. When one snapped his bike in two in 1913, he had to walk 10 miles to find a smith to solder it back together. The actual caravan of cars preceding and following the race and dropping advertisements and trinkets along the way did not appear until 1930. That same year, and until 1962, runners were required to register by nationality rather than team, which exacerbated some animosity among spectators.

Michelin, which perfected the removable inflatable tire, also initiated a new program of maps and guides to encourage riders to go further. The guides initially contained basic information intended to help excursionists find a garage to repair their tires. Soon, however, information about watering holes, hostels, and the like was added, and new initiatives such as placing road signs with mileage information began to appear as a service to bikers and automobile drivers.

The motorbike, as an extrapolation of the bicycle, elicited similar enthusiasm despite its more functional purpose. In interwar Germany, for example, the Eilenriede motor races became a focal point of both entertainment and nationalist aspirations. However, the love affair with the motorbike did not surpass that with the bicycle or the automobile. One rare exception concerned a small Italian model.

Conceived after World War II as a cheap vehicle to overcome urban distances, the Vespa continued to serve as an icon of Italian society, but one so malleable as to allow the fragmented nation to recognize itself in a variety of settings. Indeed, while by the 1980s the Vespa was marketed to young consumers as a new accessory for the well-to-do, this did not take into account the poorer sections of southern Italy that might rely on earlier models as a lifeline.

The Vespa also became, in a green-colored model 150 version, the ideal vehicle to dissect Italian society leading Nanni Moretti through the streets of Rome in his 1993 award-winning film *Caro Diario* [Dear Diary].

THE ENGINE AGE

Think of an engine as a means to turn any type of energy into a mechanical one. Until the Industrial Revolution, this usually meant wind, water, animal, and human work could create a mechanical system. Whereas the harnessing of steam had changed this equation, another engine type, the combustion one, appeared in the nineteenth century. Though early engines were notoriously unreliable, heavy, and bulky, they eventually underwent improvements that included the exact proportion of gas and air to ensure efficient fuel combustion. Variations of the combustion engine came to include the diesel fueled one, which Rudolf Diesel put together in Germany. The principle was to inject finely pulverized oil into the combustion chamber and to compress the fuel mix with air. The rise in temperature caused by the compression would bring about the burn but without the risk of uncontrolled explosion, as heavy oils do not catch fire in low temperatures. These engines, however, were very heavy and thus would become better suited to certain types of vehicles than others.

THE CAR

Early attempts at developing a roadside vehicle that would replace the horse and carriage had already appeared in the eighteenth century but were quickly displaced due to impracticality and the fact that the steam engine found an application in the railroad.

Testing of a steam engine on a road vehicle resumed in the 1870s in Europe, and it was the work of Gottlieb Daimler in Germany, who tested a 1.5 HP combustion engine, that ended all endeavors to put steam on the road. The struggle would end up being between electrical and combustion engines.

Who invented the modern automobile? In many ways, the cliché "conceived in Europe, perfected in the United States," holds true for the first half-century of the car's existence. Carl Benz patented a gas engine–powered vehicle on January 29, 1886, but Gottlieb Daimler, who became his associate, first installed an engine on a horseless carriage. Names of contemporaries of the two Germans read like a "who's who" of the automobile brand world.

Regardless of who claims what aspect of the automobile's development, the end product defines much of Europe's social history in the twentieth century. Economically first, some 10 percent of the European workforce works in a field related to the automobile industry.

At the turn of the century, the automobile remained a toy of the elite. Obviously the first to afford the vehicles were members of high society, be it gentrified city dwellers or members of the royalty. Thus, from the "ground elite" characterized by the landed nobility and industrial leadership, a new select elite appeared, this time defined by what it mastered on the road. Some members of the ruling circles appeared quite enthusiastic about the new machine, and several were pictured in their favorite vehicle as they actively promoted the cause of the horseless carriage. Prince Heinrich of Prussia, the brother of Emperor Wilhelm II, was particularly active in promoting the automobile. In fact, he provided a good balance to his brother's early reticence concerning the car, a coolness that, by 1903, had given way to an imperial dependence on the automobile. Within 10 years, as one chronicler put it, "it would be unthinkable to carry out even half of the imperial duties without the capacity to drive from city to city" ("Der Kaiser . . ." 25).

At the same time, rulers in England and Germany sponsored official clubs, races, and awards, and some even participated in such races themselves in order to encourage both drivers and constructors in their endeavors. As the winners of such races soon acquired the reputation of being larger than life, becoming as famous as the machines they drove, they signaled another important facet of the early image of the automobile: that of the race pilot.

The oddly dressed human who knew how to control the car was often portrayed as a semigod. The danger of early races, furthermore, suggested that drivers' bravado equaled that of their flying counterparts, who took to the skies in equally precarious machines and similar costumes.

Pilot portrayal was not, however, universally positive. In fact, whether a member from the ruling circles or simply a rich enthusiast, the car driver, through the excesses of his passion, risked becoming the target of sarcastic humor. Much of the humor surrounding the car concerned the aesthetic extremes to which some drivers went in attempting to match their attire to the dramatic modernity of their new machines. Among the peculiar traits the driving elite exhibited was the donning of unusual clothing, although—in their defense—drivers' outfits were designed to protect them from the variety of unpleasant conditions associated with steering early automobiles. Passengers and drivers alike had to wrestle with dust, which was already common in horse-carriage days, and also with oil and strong winds. During the urban transition from animal to machine transportation, horse dung on the streets threatened to assault innocent car riders without adequate protection. Because stopping the car might mean not being able to restart it, drivers were less than inclined to slow down to avoid such potential problems. The same went for avoiding obstacles in the car's way, which leads us to another classic theme of the new automania: breakdowns and accidents.

Writing in 1906 about automania, a German commentator remarked that the car was here to stay, and its enemies would have to adjust to its presence one way or another. To own a car before World War I meant increased social status, but it also meant exposing oneself to public ridicule, especially jokes challenging one's ability to start the vehicle, let alone control its trajectory. Happy-ending mishaps, such as an unplanned swim, joined sadder events, such as vehicles smashing into roadside trees, with an alarming frequency. In Paris in 1907, for example, a record 131 people died in the city either driving a car or as a result of being hit by one of the 6,100 private vehicles in use at the time. Advocates quickly pointed out that such numbers paled in comparison with the slower, yet more difficult to control buses and trolleys, as well as the horse carriages themselves, but the spectacle of the mechanical accident and especially its suddenness made many suspicious and even scornful of the automobile. The lack of rules to regulate the new kind of traffic was no doubt part of the problem, because courtesy was obviously insufficient to prevent close encounters of the unwanted kind. Similar misadventures also occurred outside urban areas. Although wise behavior would have called for the driver to plan a trip only in regions from which he was sure he could return, the attraction of a countryside excursion away from the city often beckoned, drawing the horseless carriage into small villages, where car owners would experience alternately a hero's welcome and the wrath of owners of livestock that the vehicle had just destroyed (see chapter 1).

Women assumed peculiar roles in the arrival of the automobile. The classic muse often appeared to accompany the car, be it as a radiator cap, a poster symbol, or, more simply, as an occupant of the machine. In Paris, aside from the society person eager to show off her latest acquisition, female taxi drivers soon appeared. Subjected to all varieties of harassment, women nevertheless remained implanted in the job until the Great War, in a manner reminiscent of the predominantly female telephone operators. Situation comedies in the form of posed photographs or even lithographs, many of which would include some form of courtship between male and female drivers or passengers, demonstrated the

This posed photograph features a woman changing a tire with seeming ease. Seeking to suggest there was nothing to the operation, the advertisement was also intended for female drivers, some of whom worked as taxi drivers. The fact that the model is not covered in dust and mud adds an air of disbelief to the advertisement. Source: Revue Suisse automobile, 1911.

empowerment of women associated with the control of an automobile. This may have been shocking for the time period, but it was also part of the magic attributed to the car: Drivers were perceived as being able to escape conventions of everyday life, including the restrictions on sexual innuendoes.

Such an accent on the female slant of technology was not all pleasant, however. Indeed, the stereotypes associated for decades with female drivers developed early and included such complaints as the wife's failure at reading a map and her husband's superior understanding of mechanics, even when the car would not start. Everybody, not just women, needed to learn to deal with the automobile.

By the 1920s, the car was here to stay. European manufacturers had begun to emulate the pioneering standardization efforts of Henry Ford. However, the best way to convince people to buy, or at least welcome the new technology was to show it in the best possible light. The first car shows took place in Paris and Berlin, but soon other shows came to dominate the scene, notably the Geneva international motor show. It soon gained a reputation well beyond local borders, taking third place in order of importance in the European Car Shows, after London and Paris, but ahead of Brussels, Amsterdam, Berlin, Glasgow, Madrid, and Turin. It is in the context of such car shows that Europeans also experienced industrial globalization first, as American cars were offered for sale. Not everybody welcomed this. A column in a Swiss magazine summarizes it best:

> America has sent us its movie stars, its bathing girls, its cowboys, its jazz bands, its bankers: it has left us with its hand me downs, it wants our money, it flood our market with automobiles, it is overstepping the mark. For example, look at the cars on the road. They all have an air of "nouveau riche", a flashy look that is not to everybody's taste. The car body is gorgeous. In it one finds all sorts of toiletries, a cigar lighter, flowers. The electrical equipment promises delicate lighting and the shock absorbers have considerations that are very touching. On the radiator cap, two symbolic wings are spread. Inside the car, a Negro mascot hangs from the window. The horn has a cheeky voice that frightens the pedestrians. All imaginable perfections have been invented. Next year, the four-wheel brakes will work even on the spare tires. To sum up, trendiness, whim and an inordinate taste for luxury are the cause of this transformation fever. (*Suisse sportive* 25)

Americanization also reached the assembly line, where Citroën was the first European manufacturer to install a Ford-inspired assembly line.

The social development of the car was accompanied by innovations in fashion. The clothes manufacturers refused to sacrifice comfort and elegance for the sake of keeping warm in the car. The clash of elite culture and driving became noticeable, for driving was a dirty business. As one columnist put it: "It is sad that some clothes make our drivers look more like unkempt fellows than the gentlemen many of them would like to appear to be" (Crane). Sports lines for car and motorbike appeared, and designers competed with each other by cutting coats with leather sleeves serving as both windbreakers and mittens, as well as foldout coats with a gimmick to transform these temporarily into full body suits.

Racing

During the Paris Car Show of 1922, Emile Coquille, a tire manufacturer, suggested the idea of a long endurance race for cars on main roads. This project came to life the following year. The first Le Mans 24 Hour Race took place on May 26 and 27, 1923. It involved racing on national and departmental roads, a good opportunity for putting to the test the headlights of the vehicles at night. No ranking as such was planned, but the competitors had to cover a preset distance varying according to the power of their engines to qualify for the race again the following year. Attracted by this first event, the British came back in force in 1924, and this time the French cars were defeated by Bentley.

Le Mans quickly gained notoriety, but so did Monaco. The Principality's Grand Prix had been held in its streets since 1929. As one magazine described it, the show was as much in the street arena as it was among the spectators:

> [It is] an occasion worthy of attracting lovers of strong emotions seeking unusual events. There is no more surprising a sight than the one offered by this city, so protective of the beauty of its walkways and the nonchalance of its hosts, when it is abruptly awoken from its torpor by the engine roar of more than a dozen monsters encircling the luxury hotels, piers and the palace in an infernal merry-go-round. . . . The avenues are turned into dangerous tracks with each bend, each palm tree becoming a deadly obstacle. The inhabitants have deserted the streets; doors are bolted by order of the police. One stays away from the madmen. Spectators take refuge on balconies and terraces from which they hang like grapes. The elegant ladies who were betting on horses the night before are now betting on cars. There are yells in every language when the racing drivers representing participating countries race by: this is where France, Germany and Italy have placed all their hopes. (Monaco 7)

The race mania increased after World War II. Every large city in Europe and the United States wanted its own car racing event. An international commission made up of several sport officials attempted to clean up the prevailing anarchy so that in the early 1950s, an agenda and regulations were established for several technical formulas, including Formula number 1, known today as F1. The International Automobile Federation instituted a world driver's championship to be held over six grand prix spread throughout Europe (Monaco, Belgium, England, France, Switzerland, Italy) and Indianapolis. The number of grand prix and their locations evolved through the years in response to traffic, cost, and pollution regulations, some of which banned the races (Swiss grand prix).

Traffic Rules

Paradoxically, learning to accept the automobile involved not only educating drivers but also pedestrians. It wasn't generally until the 1930s that

Patent-Schutzvorrichtung.

"Patented protection." The caricature pokes fun at the lack of socialization both on the part of the driver (who knows no brakes) and the pedestrians (who do not get out of his way). Source: Astra, Geneva, Switzerland; reproduced by permission.

formal legal rules were enacted in most European nations. These rules required not only insurance and maintenance of the vehicle but also proper behavior on the road founded on a kind of ethics of mutual responsibility (Moran 479). The reason for such codes included the high death rate on roads, which exceeded well over ten thousand a year in most European nations.

The European Traffic Congress, held in Geneva during the Spring of 1931, resulted in the convention of April 10, signed by Germany, Belgium, Denmark, the free city of Danzig, France, Italy, Luxembourg, Poland, Switzerland, Czechoslovakia, and Yugoslavia. It established a joint system of road signs and markings divided into seven sections: seven triangular danger signs to signify a stop or an obstacle; red discs barring access to a street or passage; disc-shaped signals that indicate the direction of traffic with an arrow on a blue background; caution signs with a triangle on a square background; simple indication panels such as blue rectangles pointing to first aid stations or parking lots; and panels showing the direction to or the name of a locality.

It was during this time that automatic signaling was instituted, less costly and cumbersome than traffic officers at intersections, while researchers were busy designing shiny or luminous pedestrian crossings as well as rubber bollards.

The 1920s and 1930s also saw the beginnings of the small personal car in Europe. In France, Citroën produced a prototype, the 2 CV, intended as the most affordable and easiest to use. It would enter mass production

A pre–World War I practice gained further ground in the 1920s as open bus tours were offered to tourists in many cities. Here, a newly arrived group has exited the Cologne train station next to the cathedral and boarded an open bus. Source: Private photograph, author's collection.

after World War II. "The duck," as German youth called it, was popular because it was cheap and practical. Initially, though, its success was such that someone placing an order for it in 1950 could expect it in . . . 1955!

The Beetle

The *Volkswagen* or "people's car," was never named thus in its original inception. The brainchild of Ferdinand Porsche, it became a personal fascination of Adolf Hitler, who had developed an early fascination for automobiles and used them extensively in his campaigns. Hitler had ordered special tax breaks for car licenses and initiated early on the building of the autobahn. What fascinated him in Porsche's proposal was the notion that a modern car could reflect the ideals of the new Nazi German community. The *Kraft-durch-Freude Wagen*, as it was initially called (and it retained that name up to World War II), would allow for countrywide tourism at affordable prices. The reality was otherwise. Workers who agreed to set aside some funds every month would, or so the advertisements argued, eventually be able to place an order for the car, valued at 990 reichsmarks. In reality, wages and upkeep (estimated at 70 reichsmarks due to high gas prices) meant that only the middle class could really afford such a car. By the time the war began, not a single civilian version of the car had been delivered. In war, some 340,000 were made, many in the *Kübelwagen* (jeep) and *Schwimmwagen* (amphibious) versions; this was a far cry from the projected 1.5 million vehicles.

After World War II, with assistance from the British occupation forces, a factory was reopened at Wolfsburg, and the VW 1100 began to roll out. The beetle ended up joining a series of smaller cars designed around World War II but manufactured into the 1950s that allowed more Europeans to own a car. The small size was attractive in cities where traffic and building patterns limited the amount of parking space. Though most of the Beetle's models were notorious for being underpowered, it was appreciated in urban areas for its sturdiness, and it became the workhorse of many police and public works departments for up to three decades. It did not, however, acquire the image of youth rebellion associated with the models sold in the United States.

By the late 1950s, smaller vehicles had found a niche among cost-conscious elements of the population as well as the well-to-do youth growing up in a more comfortable economic. In France, the Citroën 2CV was *the* fun car, or so the manufacturer's advertisements suggested. Cheap, flexible, and with a unique shape, it suggested mischief and relaxation rather than austere cost-cutting measures associated with the 1930s, when it was first conceived. First presented in 1948, it remained in production until 1990. Though staggeringly successful, its iconic power was somewhat eclipsed by its contemporary German counterpart, the "Beetle."

The aftermath of World War II left Europe lacking in everything: rubber, gasoline, and labor. The French steel reserves were 40 percent below those of 1938, and their allocation to the automobile industry was completely insufficient. All that was left of Italian factories was what had survived the bombings. In England the coal shortage caused the closure of many plants during the winter of 1946–1947. The bombardments had destroyed a number of firms: Daimler in England, Mercedes-Benz in Germany, Peugeot and Renault in France, and Lancia in Italy.

Vehicles were scarce; when they were available they had no tires. The number of passenger cars in France was estimated at two million, but barely half of these were operational. In Germany, transportation had come to a complete standstill. Switzerland had been spared from combat, but it badly lacked in transportation facilities. Supply sources were severely limited. Both from an economic and technical point of view, America was several years ahead of Europe, but both sizes and prices of its cars were too great for Europe—a Chevrolet stretched to 15 feet, and its gas consumption was too great for lean times.

In the first decade after World War II, cars remained expensive but became necessities. French manufacturer Renault's 4CV model, a very popular model, cost on average 30 months' worth of a working-class salary (by contrast, an equivalent model in the 1990s would cost about 7 months' salary). The success of the popular car culminated in the Beetle but also in

Car as leisure. With the improving European economy in the 1950s and 1960s, more Europeans took to using their cars to go on leisure trips, in this case up to a ski resort in the Alps. Source: Astra, Geneva, Switzerland; reproduced by permission.

British Leyland's Mini, a diminutive car that also announced the transition to performance vehicles in the 1960s.

The 1960s saw a car industry in full bloom in Europe but facing a few clouds, notably in the realm of safety. New traffic rules went into effect to try and reduce the number of accidents (such as no crossing of double lines). It was an essential move, for the 1960s also saw the increased development of highway links in response to both postwar recovery and the increase in number of cars on the road. Automobile clubs iterated wish lists for safer cars, which included such things as: safety belts at least for the front seats, a steering wheel and column safe enough to sustain a collision, tires built for speed and weight, accessible controls, at least one external side mirror, windshield wipers, security locks, elimination of external or interior sharp edges, adjustable seats, lockable seat backs, efficient ventilation and heating systems, and a braking system designed to prevent premature locking of the back wheels.

With the increased exposure of Europeans to Anglo-Saxon popular culture, the automobile also appeared in highly popular movies. It thus extended the tradition of desire for a speed technology. The automobile forsakes reality for a fantasy world by combining all its assets with those of power and eroticism, beginning with James Bond 007 movies in 1962 and the first Pirelli calendar in 1964. In the Bond movies, the star cars featured included product placements in the form of cars, such as the Sunbeam

Alpine (*Dr. No*), Aston Martin DB5 (*Goldfinger, On Her Majesty's Secret Service*), Toyota (*You Only Live Twice*), Hornet, Lotus, and BMW. Other films of the period, such as Dennis Hopper's *Easy Rider* (1969) or John Schlesinger's *Midnight Cowboy* (1969), portrayed the car or motorbike as a symbol of freedom or, conversely, as a nightmare in Jean-Luc Godard's *Week-End* (1967).

The end of the honeymoon with powerful vehicles came in 1973 with the first oil shock. The massive jump in oil prices prompted Europeans to rediscover diesel. Already used in commercial vehicles and by Peugeot in the 1960s, though costly to produce, this engine is very economical to use in any type of car. Following Peugeot and Renault, Mercedes-Benz, Volkswagen, Ford, Audi, and Opel began to equip their cars with diesel engines, although the American car industry remained totally recalcitrant. Instead of drawing in gas, the diesel engine draws air into the cylinders and then compresses it in order to reach the right temperature for the high-pressure injected fuel to ignite independently, that is to say, without the need of an electric spark. This concept was discovered by Rudolph Diesel. The engine capacity is modest, but the fuel is much cheaper than regular gasoline and is particularly suited to cars covering long distances. Already in development in 1929 but fine-tuned after World War II, the diesel engine became very popular in the 1970s due to its reduced consumption and its adherence to gas emission norms. The power is improved, but there is still work to be done on noise, vibration, weight, smell, and especially cost. The extra expense of a diesel engine can only be offset after a substantial number of miles on the road.

With the contraction of the world economy, manufacturers sought new ways to sell their production, including through cross-marketing similar models rather than importing or building new ones. This also allowed some manufacturers to avoid strikes, social problems, and economic crises in their own countries. Ford relocated some of its production to Germany, Spain, and Great Britain. In 1977, Chrysler produced its first international car under various names: the Simca Horizon in Europe became the Plymouth Horizon in the United States. General Motors had been manufacturing its Chevrolet Chevette, twin-sister to the German-built Opel-Kadett, since 1975. Finally, the 1981 European-made Opel Ascona and Vauxhall Cavalier are identical to the Chevrolet Cavalier, the Pontiac Sunbird, the Oldsmobile Firenza, and the Buick Skyhawk.

Globalization was already in full swing, and the increase of Japanese exports to Europe as well as the arrival on the scene of Korean manufacturers offered Europeans more choices on the car market and more headaches in the balance of trade.

With mounting consumer concerns for road safety, pollution, and energy consumption, numerous environmental movements emerged. Addressing these concerns, car producers began extensive research programs on braking systems, visibility, and steering, and politicians pledged to improve road conditions, lighting, and signaling. Swiss law, for example, requires every person applying for a driver's license to take a short first-aid course to assist the injured in case of a car accident.

In 1978, two public initiatives put to a vote by environmentalists were rejected: one for "more democracy in road building," the other named "Albatross" for swifter measures against noise and pollution. Another public consultation called for "Car-Free Sundays." All were defeated. As in America, Europeans depended heavily on their motor vehicles despite the presence of very good transportation networks.

The second oil shock in 1979 heralded the triumph of the small car in the 1980s, though by the mid-decade, with the European economy improving, Europeans looked askance at a new model presented at car shows: the minivan. A new accessory also became available to the fortunate: a car phone that was easy to install and operate. In both cases, these novelties remained that for almost a decade, with the former catching on in the United States, and the latter developing faster in Europe.

TRUCKS

One of the immediate effects of the flourishing of car constructors in the early 1900s was the development of the utilitarian machine, be it a bus or a delivery truck. Originally, people had believed it sufficient to simply transform the car frame to create larger vehicles for transport. Soon, however, and in conjunction with the design transformations that accompanied the development of the civilian car, sturdier chassis without the contraptions that equipped luxury cars began to appear. Their lower cost made such vehicles attractive to salespeople and city officials alike, and many constructors recognized the possibility of surviving financially by catering to such clientele in the face of high competition (similar reasoning apparently influenced the auto-support business, particularly the tire industry).

The utility of trucks and buses was quickly demonstrated in most European cities. Though some already appeared prior to World War I, it was developments in the conflict that accelerated the devising of new uses. Some were reminiscent of those found in the United States, though with variations: library buses and advertising vehicles come to mind. Consequently, a new phenomenon began to appear by which the advertising potential of the automobile was exploited so that the real machine started carrying advertisements. Delivery trucks began to sport prominently the brand of the manufactured goods they carried, and as one writer pointed out, this was no longer a case of the car for car's (or transport's) sake, but rather, "the business vehicle is . . . a valuable complement to the propaganda which the modern businessman uses to remind everyone of his name. The business truck is a 'wandering advertising column'" (Kirchner 60).

It was not just the truck itself, though, that fascinated. There appeared a driver culture associated with trucking. Though it did not resemble

Trucks were put to all kinds of uses. Here, a modified open-carriage model from 1912 is used to move paralyzed pilgrims coming to Lourdes, France, to pray for their recovery. Source: Tourism postcard, author's collection.

that in existence in the United States (where distances alone account for a major difference), there was nonetheless a tradition of trucking present on European highways. In the case of Great Britain, the truck (or *lorry* as it is called in queen's English) was part of the driver's identity. This was due to a combination of factors that included the nationalization of all trucking businesses for a few years following World War II; the specificity of models developed for specific industries; and the complexity associated with driving haulers. Furthermore, drivers were even expected to help load and unload their cargo regardless of the long-term health implications. By the late 1960s, technological improvements as well as changes in licensing of drivers meant that the culture associated with postwar trucking disappeared in favor of a more anonymous culture, where drivers were not associated with a particular type of truck (Cooper 69–81).

In some cases, the trucks themselves acquired an identity all of their own. In Switzerland, through its grocery trucks, an entire company became synonymous with food supplies. In 1925, Swiss businessman Gottlieb Duttweiler turned an idea into reality: Frustrated by the high prices of groceries and other items of first necessity, he wondered whether it would not make more sense to have a single clearinghouse that would bring consumers the needed products without going through intermediaries. Duttweiler founded the Migros company. In August of that year, the first five trucks of the corporation, using Ford Model Ts transformed into wheeled showcases, began traveling around Zurich and stopping on demand to sell anything passersby saw on board. The infrastructure, kept at a minimum, consisted of an office, a shop that managed deliveries, and a coffee bean depot. The success of the company quickly resulted in the creation of associated regional cooperatives (where farmers brought their goods) and the spread of the food truck service to several hundred cities and small villages around Switzerland. The country's topography made many remotely accessible habitats hard to service for permanent stores. The assurance of a truck visiting regularly relieved many women, traditionally in charge of the household, from having to pool resources to send someone to the city to fetch a long list of goods.

Competition actually forced Migros to diversify even further by producing some of its own goods, yet in so doing, it also kept low prices and attracted many more clients in the rural and working-class sectors of the population. Of course, the need to serve greater areas meant a diversification of the trucks. Ford Model As soon replaced the old Model Ts; Chevrolets soon appeared, and by World War II, locally built trucks, like Saurer, also represented Migros on the road. A self-service truck was also introduced in 1960.

Enthusiastic about its success, the company tried several foreign developments. In Berlin, a subsidiary was successful until the arrival of the Nazis in 1933. Other countries, mostly less developed and rural, drew the interest of Migros. In Turkey in the 1950s, a new operation using British trucks was very successful and eventually spun off on its own by the mid-1970s. In Spain, however, a similar attempt in the 1960s failed.

The increased mobility of individuals also reduced the need for grocery trucks. Migros's fleet is considerably reduced nowadays, as it only serves a few Swiss mountain villages. The company is now mostly associated with the chain of grocery stores it began inaugurating in the 1930s that still run today. What the Migros experience showed, however, was that using trucks as mobile grocery centers was the right solution at the time and made effective use of truck technology.

CONCLUSION

From a dream toy for the rich, the European car became, like its American counterpart, a central necessity. Changes in living patterns mean nowadays that urban planning practices had to account for a generally different segment of the population commuting inward to the city center to begin jobs. Whereas many factories were of course located in the suburbs, many working-class jobs were also to be found downtown. Consequently, it became essential to have the means to reach the workplace efficiently. Cars used to be out of the question, and public transportation did the trick. Yet, commuting drivers have increased in numbers. The convenience of not being held to specific departure times and travel at will seem to outweigh the problems of traffic jams and difficult parking for many.

By the 1990s, road congestion, a dearth of parking spaces, and increasing noise and pollution were present everywhere. In spite of Europe's efforts to curtail city driving, California's advocacy of car sharing, Mexico's alternate "even- and odd-day driving," and the concurrent promotion of public transportation, the problem remains unabated. This does not prevent manufacturers from producing smaller cars designed for city driving. To alleviate traffic congestions, an old idea was reintroduced: the roundabout. The very first one was inaugurated in Paris in 1907, Place de l'Etoile. The principle involves no traffic signals; by forcing cars to turn around an obstacle—the Arc de Triomphe, in this case—and all move in the same direction, traffic jams are avoided at crossroads linking up several avenues. Many reasons lay behind the success of the roundabout: safety, fluidity, and reduction in the number of accidents thanks to a steady flow of traffic and moderate speed.

5

NAVIGATING THE EARTH AND HEAVENLY SEAS

A third group of machines that changed the way Europeans understood their everyday life included the boat and the airplane. The coupling of these two modes is not as incongruous as it may appear at first. While the boat in its hundreds of functions is almost as old as humanity, its transformation through the use of steam and other engines into massive freighters and ocean liners completed Europe's link to its colonies and accompanied millions of emigrants around the globe. The so-called queens of the seas became surrogate symbols of a country's power and prestige, and in so doing combined the public's fascination with speed with the mass transportation possibilities the railroad had brought about. The airplane did the same and ended up displacing the sea liner on many routes. Yet, as a child of the twentieth century, the flying machine's impact, though immense, took place in different stages.

BOATS

Perhaps the most distinguishable feature of the European experience on the seas was both the increasing diversity of ships and the combination of size, splendor, and luxury associated with one particular type: the ocean liner. Associated with the late nineteenth-century royalty that often partook in inaugural launchings of such giants, these machines remained on the forefront of long-distance travel until the airplane took over in the 1950s.

The modern era began squarely with the use of sail and witnessed major battles fought the way they had been for centuries. Fleet commanders such as Admiral Nelson also acquired a heroic status that inflamed the popular imagination of the British public (Nelson defeated the French Napoleonic fleet twice). But the technology that had given Europeans the means to colonize and dominate other nations was about to undergo a new revolution, through the combination of the steam engine and the iron hull.

The tradition of royal sea launches started with Prince Albert's involvement in the christening of the *Great Britain* in 1843, using the customary bottle of champagne. The ritual celebration was good business for shipping companies and became the subject of news coverage, especially when celebrities were present. *Great Britain* was also unique in that her designer, Isambard Brunel, had decided to apply the revolutionary concept of fashioning the hull out of metal and using propeller screws for the engines instead of paddles. Brunel had also built the *Great Western*, which was the first steamship in regular transatlantic service.

The *Great Britain* ran aground the following year, but because it had been designed with an iron hull, it was repaired and sailed anew. This success

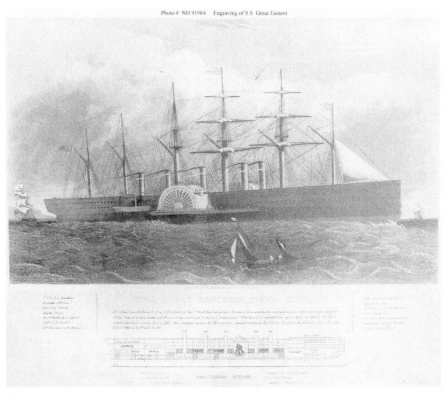

The *Great Eastern* as it appeared to contemporaries. It set the stage for the design of great ocean liners. Because engines were still a novelty, ships still were built with masts and sails in the event of machine failure. Source: Naval Historical Center.

inspired her designer to build an even greater ship, the *Leviathan*. The ship, eventually renamed the *Great Eastern*, was a financial and commercial failure, but it stuck in the minds of contemporaries for its size. It was some 680 feet long and could displace up to 30,000 tons. The technology was available to build something this huge (after her, there would not be another ship this big until 1901), but there were many failures at refining it. Steam pressure was wasted, the ship tended to roll in heavy seas despite its mass, and few ports could actually accept something this monstrous. The one success associated with it, however, was the laying of the transatlantic cable line, which allowed the transmission of telegraphic signals.

The commercial failure of the *Great Eastern* nonetheless set the tone for bigger ships, and the days of the sail had disappeared by the 1890s. Steamers became common on lakes, rivers, and seas. First as paddle boats and later as screw-propelled, such ships projected an aura of safety. They also offered a genuine service as shuttle ferries, connecting travelers across the water to railroads and roads. Nowadays, ferry services that still exist often are the result of a collaborative enterprise involving rail service for greater efficiency. Some of these services, however, only turn a profit thanks to nostalgic tourism.

The other impact of the increase in size was the possibility of greater trade. Sail ships had already reached an impasse in the eighteenth century,

A ferry service between Dover, United Kingdom, and Calais, France, unloads private cars in the 1930s. The ferry service continues nowadays, despite some strong competition from the Eurostar train that tunnels under the Channel. Source: Astra, Geneva, Switzerland; reproduced by permission.

Among the hundreds of ship designs drawn and built in the late nineteenth century, several involved cable-laying designs for intercontinental communication. Here, the *Faraday*, a German ship, awaits its next mission. Source: Courtesy Siemens Pressebild.

whereby one could not increase their size any further. With the iron hull, later made of steel, the problem was swept aside. The first oil tanker, for example, was the British-built *Gluckhauf* in 1886, which stored oil directly in the hull. By 1900, most oil tankers were in European fleets, but after World War I, U.S. interests would take the lead as energy consumptions shifted from coal to oil.

Hundreds of other designs existed, yet, these did not always serve their expected goal. The British empire, for example, to ensure control of its territories, relied on river patrols and passenger boats. The design of traditional European steamers, however, implied the ships drew 8 to 12 feet in the water. This was far too deep for many rivers, and colonial authorities made numerous demands for the design of shallow-draft steamers that would ensure swift upriver and coastline communication (Kubicek 427). In the case of designing boats for imperial use, though, notable tensions arose between what designers and bureaucrats in London offered and what local authorities expected. The result was that the foremost expert in high seas gunship design, Sir Edward J. Reed, proved unable to account effectively for operating conditions in the tropics and did not redesign ships until official pressure came to bear on him. Meanwhile, local colonial authorities contracted directly with local British shipbuilders instead to obtain more efficient boats. In the end, though, a design begun in the 1860s took almost four decades to mature (Kubicek 449).

Though the size of steamers increased in the twentieth century, working conditions remained fairly similar to those encountered earlier, as many boats still used coal rather than diesel for their engines. The thousands of horsepower the engine developed required ant's work in the engine room, where a journalist tried his luck at shoving coal in exchange for free passage:

> A manhole, three steps down. Here I am in the engine room. It's huge, lit by daylight. In the middle, a grill-enclosed steel monster. . . . Three men come and go with oiler in hand, careful, attentive. Sometimes checking with caution a small piece. They are the mechanics. I am just a stoker. I will soon see the difference. . . . Let's go! Says the chief stoker, used to ordering a multilingual team in English. He points to a black hole at the other end, sometimes lit by yellowish tones. It's the entrance. I enter the dark tunnel and feel suddenly the burning air, loaded with coal dust. . . . Temperature: 55 degrees Celsius [131°F]. A dangling electric lamp throws light onto five men with tense faces. Sometimes, one grabs some rags and lifts a lock to pull a double door. Flames jump out and illuminate the sweat and coal-covered faces. In a corner, a pale of fresh water constantly renewed has a few quarts dipped in it. One drinks there in a rush, between two reloads, for the fires do not wait. Fifteen minutes of slowing down and the pressure will drop. Next door, the chief mechanic checks the dial constantly. It cannot go below 90. That's the business of the stocker and the two storemen who bring wheelbarrows of crushed coal. . . . I must, like the others, feed two of the twelve fires. I take my place, rags and shovel in hand. I open one of the two fire doors and stuff everything in fast. It's not difficult. You just have to spread the coals and mind the rising flames. Strange, the rate of the wheelbarrows sees to rise. After fifteen minutes, I feel like I won't hold more than a few seconds. Then, I get used to it. (Hamre 251)

This glimpse of working below decks was of course further masked by the advent of the ocean liners. Whereas the experience of sailing the sea is almost as old as human civilization, the practice of doing so in a floating town is quintessentially an experience of the late nineteenth and early twentieth centuries. Its representation to the public usually involves the best part of such liners, ignoring the majority of passengers in other sections of the ship.

The pattern is essentially the same no matter which company one considers: A few hundred wealthy clients would be treated to the ultimate experience in the upper realms of the ship, while below deck as many as 2,000 souls traveling steerage (later renamed third class) would undertake the immigration experience, primarily to the United States. The Hamburg

Amerika Line, for example, offered twice- or thrice-weekly departures, and British and Italian counterparts did the same.

It is in this context that the tragedy that befell the White Star Line's *Titanic* in April 1912 had a massive impact. The ship hit an iceberg and sank, but most of the survivors were from first-class areas. The resulting angry discussions that followed in both the United States and the United Kingdom presaged in some ways the end of the nineteenth century and the advent of a new social order. Years later, one of *Titanic's* sister ships, the *Britannic*, was torpedoed and sunk. The other sister ship, *Olympic*, survived the war and was used until she was sold for scrap in 1935.

At the level of routine, it is worth pointing out that boats served colonial destinations with the precision of railroads on land. Though not as luxurious (or profitable) as the North Atlantic route, the Peninsular and Oriental (P&O) service was subsidized by the British government to bring mail, supplies, soldiers, and even women principally to India. The people who worked these lines were not just British but an amalgam of Europeans and South Asians. Crossing the Suez Canal and then the Red Sea was the slowest part of the trip and the hottest. Passengers, when they booked, were advised to ask for *port-outward-starboard-home* accommodations so that they would not be on the side of the ship most exposed to the sun. The expression's acronym (POSH), used by booking clerks, has since become a regular word describing the privileged (Woodman 257).

Once the element of adventure associated with departure dissipated, boredom set in, as noted in the journal of a functionary's young wife on her way to join her husband in Indochina. This mother of an infant emphasizes the balancing act between fascination at the places visited and sadness as she struggles to care for her child alone. After a few days at sea, the element of novelty is lost, and one must devise a new routine that does not include shopping, gardening, or even reading one's mail: Failing to become accustomed might cause depression (Mechin and Mechin).

As for the Atlantic route, the ultimate in the combination of luxury, speed, and prestige was likely reached in the 1930s, despite the economic depression. The French government offered a $60 million subsidy to help build the *Normandie*, which became by most standards the epitome of taste. It claimed the blue ribband soon after entering service in 1935 and competed directly with the British *Queen Mary* and the German *Bremen*. As far as luxury went, the *Normandie's* was unequaled, and a design style associated with it (French) became widespread. The end of the ship was all the more sad that it had been stored in New York City when World War II broke out. A fire set accidentally spread, but the water the firefighters doused on it forced the ship to keel over. Though raised, she was eventually scrapped.

After World War II, ship building resumed, but its associated costs and the need to rebuild battered European economies meant smaller liners would appear. The fact that the airplane was quickly gaining in popularity also signaled the end of an era. The last time the blue ribband was awarded, the American S.S. *United States* earned it, unchallenged. A latecomer, the S.S. *France*, which sailed in the early 1960s, was the last great ocean liner designed for that purpose. Its owners bankruptcy in the mid-1970s led to its sale and reconditioning into a ship for Caribbean cruises.

The demise of the *France* while a few other ships carried on in a more restricted market confirmed that the trend was now toward leisure travel and less on prestige. Boats were now designed to accept entire families with varying tastes. The first-class experience was quasi universal, with only the size of the private cabins marking how much a person had paid. Niche markets have taken over but are common everywhere. True, some ferry services between countries (such as Sweden and Norway) are prized for their duty-free shopping opportunities. Travelers take advantage of the fact that they are in international waters to enjoy bouts of drinking but also gambling at onboard casinos. Yet, this can happen anywhere, from the North Sea to the Bermuda triangle. Globalization has taken over sea travel.

The Flying Machine

The European interest in flying can be traced back to the nineteenth century, and several pioneers of that era are nowadays honored as precursors, even when their own flight endeavor failed. Two kinds of flying fascinated Europeans: lighter-than-air and heavier-than-air. The former predates the French Revolution, whereas the latter accompanied the twentieth century. In both cases, the fascination mixed with fear and awe.

As far as lighter-than-air (LTA) contraptions go, the balloon was the first incarnation, with a 1783 flight in France. In the French Revolution, a tethered balloon was used at the battle of Fleurus to observe enemy troops. Other than that, however, balloons came to be viewed as means of entertainment with little other purpose. By midcentury, tethered balloons were often installed at fairs and expositions. Circuses, on occasion, featured an acrobat suspended to one.

Scientists first saw a potential for ballooning, as some sought to take atmospheric measurements. In France, the artist Nadar commissioned a giant balloon to take him and his bride on their honeymoon. A crash followed, but their survival made him famous. The problem of ballooning was that one knew where the contraption took off but never where it landed. Attempts at solving the directional challenge generally failed, be they on paper or in practice. By the late nineteenth century, LTA was often associated with *balloonacy*, a term coined in Great Britain to describe a field of inquiry that combined bona fide science with complete fantasy.

Hence, when dirigibles first appeared in the late nineteenth century, they failed to convince both the public and the authorities that there was any value to them. Too slow, the wind swept them as much as it did balloons. Filled with hydrogen, their accidental destruction tended to reinforce the conservative attitude that God never meant for men to fly.

A humoristic depiction of the explorer Andrée's arrival at the North Pole. The optimistic tone masks what was already assumed to be his death aboard the balloon. That technology saw a rebirth in the late nineteenth century but was eventually replaced by airships and airplanes. Source: Postcard, author's collection.

Explorers relished using them but were far from successful. Arctic explorer Andrée for example, set off aboard the Eagle in 1897. He and his companions crash landed there, and they survived into the winter. Their bodies were found some thirty years later. In the meantime, their disappearance had given grounds to random speculation, from fantasy to horror.

This is why the invention of Count Ferdinand von Zeppelin eventually became such a landmark in aviation history. Patented in 1895, the first model of Zeppelin's conception flew three times in summer 1900 before being taken apart. The press covered it, and people took an interest in the attempt, but only as a form of entertainment in the hot summer months. It would take Zeppelin another eight years before achieving consecration with his fourth machine, which flew a long-distance record in August 1908 before crashing. The accident, which caused no injuries, was perceived as an injustice by the German public, and millions of marks poured in within days to help the Count rebuild his machine. The event itself symbolizes many facets of the public fascination with aviation, in Germany and throughout Europe. As they became fascinated with dirigibles, Europeans (not just Germans) projected a future where airships would act as trains or buses do, while airplanes would be like private cars.

Thus, the world's first air carrier, DELAG, was established to offer rides aboard Zeppelins. These were circular trips, intended to amortize the cost of maintaining a fleet of unsold machines (the military remained cautious about the airship). Yet, by 1913, hundreds of well-to-do passengers had experienced the unique stability of an airship cruise in fine weather. The German military began acquiring them, although it was unclear how useful these might actually be in war (military maneuvers with them had proven inconclusive, despite some promising potential as observation platforms). The war interrupted plans for further passenger expansion.

After 1918, the Allies actually sought to appropriate German dirigible designs as a means to achieve a military advantage. Indeed, German airships, though constituting what is called a "peripheral weapon," nonetheless had a substantial impact on populations by becoming the first strategic bombers. That more were lost to weather incidents than to actual combat was overlooked: Every Allied nation wanted its own airship fleet. Yet, the public fascination with these remained limited, even though one, the British R 34, first crossed the Atlantic in 1919.

It would take another German effort to reestablish an airship craze in Europe. Count Zeppelin had died in 1917, and one of his replacements at the head of the company, Hugo Eckener, succeeded in saving the airship branch of the company (it had diversified into aluminum products) by first delivering an airship to the U.S. Navy in October 1924. The ZR III, later christened Los Angeles, proved popular to newsreaders across Europe. Using this success, Eckener parlayed funding to construct a prototype airship intended to demonstrate the value of long-distance flights, notably across the Atlantic. The LZ 127, called *Graf Zeppelin*, would become the most successful airship of all time. Like its eventual successor, the *Hindenburg*, it carried passengers and mail, and by the 1930s was on a scheduled rotation across the South Atlantic. The *Hindenburg* did the same between Germany and New York in the 1936 season (airships never flew in the winter months).

Airships seemed attractive to Europeans and Americans alike despite the risk of explosion (the only helium reserves were in Texas and were held as strategic material that could not be exported without a permit). The ride was smooth, there were few air bumps, and the experience, in the words of the happy few who could afford a ticket, was akin to that aboard the first class of an ocean liner, only much faster. The British tried their luck with two designs, including the R101, which crashed and burned on its maiden flight to India from England in October 1930. This left Germany alone in the passenger business. The *Hindenburg* promised to reach New York in 2.5 days, whereas the fastest ships, such as the German *Bremen* or the French *Normandie,* required at least four between the continents. Only flying boats seemed to be likely competitors.

The destruction of the *Hindenburg* on May 6, 1937, marked the end of the airship era. Since then, a few blimps have graced the skies of Europe but generally for advertising purposes only. A recent endeavor by the re-constituted Zeppelin airship division to offer tourism flights above certain regions has proven extremely successful, but this is a niche market, unlikely to grow to mass proportion. Though LTA is no longer a principal means of flight, its demise should not be taken as a given. Until the mid-1930s, it appeared that, unless speed and endurance problems were solved, the airship would remain sovereign in the skies when it came to long-distance flights. Nowadays several European companies have sought to bring back the airship as a heavy cargo carrier, but these efforts have, thus far, failed. The primary reason is that the required infrastructure (airports, docking areas, etc.) are designed to service airplanes.

Although developed in quasi-parallel with the automobile, the airplane did not enjoy the same degree of socialization as did the former. Whereas the car soon became a heavily produced item relatively easy to use, the fly-ing machine, while fulfilling an almost eternal human dream, inspired as much fear as it did enthusiasm. Not only was air travel exclusively for the wealthy, but—from the viewpoint of many—it was also only for the fool-hardy. For example, Louis Blériot, an aircraft constructor seeking funding, was once told, "I'd rather you drank than built these airplanes."

Yet, acceptance of the airplane came in part through news propagation by the press and other means of popular representation. The first post-cards depicting flight themes represented balloons such as those flown at world expositions and dirigibles. Other shots soon followed, particularly of glider experiments, from Lilienthal in Germany to Captain Ferber in France. Heavier-than-air machines did not really follow until the news of the success of the Wright brothers reached Europe. Soon after, various constructors, serious and not so serious, started working on solutions to the flight problem.

Generally, though, the Wrights and their models dominated the scene thanks to the ease with which it could be steered. Time and again, the Wright machine appeared at airshows, here flown by Count Lambert in France, or even by C. S. Rolls in England, where the latter was as inter-ested in automobiles as he was in airplanes. The effect of such representa-tion was to suggest a kind of brotherhood of aviation in which nationalist boundaries had not yet been set. A Frenchman taking up an American ma-chine did not raise questions of national betrayal. One card, for example, depicted French aviation pioneer Tissandier receiving information from Wilbur Wright, while another postcard issued in the series "Our Fliers" portrayed Wilbur as one of the crazy Gaullic flying troop. It wasn't in fact until 1910–1911 that nationalist tendencies took hold, as European gov-ernments opened public subscriptions to form army air wings. Until then, however, the image of the airplane, including the Wright flyer, communi-cated a sense of universal victory over the elements.

With World War I, the airplane shed its image of innocence, but technical progress inspired the establishment of the first air transport contractors.

Airshows became the first contact between the general public and aviation, and stories associated with flying were very popular. Here, the cover of an Italian magazine depicts an air meet in 1910. Source: Author's collection.

The primary interest then was the carriage of mail over long distances. Air companies, using modified World War I aircraft, would receive government fees, which, combined with income from private mail, would allow them to operate various lines. The experiment worked but did require governmental intervention. Too many small private carriers existed, and the resulting cot was exorbitant. This is why national flag carriers appeared in the interwar years. Some of the oldest ones still in operation include the Dutch KLM, Air France, and Iberia. Others, like Lufthansa, British Airways, or even Swiss (formerly Swissair), derive their existence from earlier incarnations but are separate corporate bodies.

In the interwar years, flying as a passenger was not just a matter of prestige but of adventure. Winter schedules did not appear until the 1930s, and even then the issue of comfort was foremost in the minds of consumers. For a higher price than cost a railway compartment or a boat cabin, one was entitled to a wicker chair with a belt, a loud engine noise, vibrations, and even the occasional whiff of hot oil or burning fuel coming from the neighboring engine. The latter sometimes quit, too, and emergency landings in the countryside were frequent. To attract clients nonetheless, airlines resorted to serving fine foods (initially cold) and liquors. The advent of flight attendants, first men, was simply to ensure there was no panic in the cabin while the pilot(s) flew the machine. By the time female flight attendants appeared (the first one, Nelly Diener, flew with Swissair in 1934 but later died in a plane crash), their purpose had become to reassure passengers by providing them with blankets, aspirin, and to help clean up any vomit. This bleak picture should nonetheless acknowledge that progress did occur, but it relied on technological improvement.

The interwar years also saw a special fascination with famous figures of aviation. Prewar, a few major figures had cut imposingly on the media canvas. After the war, itself fed by the image of the heroic aviator, Europeans came to admire those who performed masteries of technology. Ernst Udet in Germany fascinated with his stunts. France admired Costes and Bellonte, long-distance record holders. Italy feasted Balbo, and Portugal Admiral Couthino.

Because aviation had entered this transitory phase between experiment and routine, it retained a magical aura akin to that of balloons and circuses. Several women saw an opening and tried to work as pilots, either as stunt fliers or record breakers (the difference was often erased because sponsors wanted air show demonstrations). An international convention banned them from becoming commercial pilots, and this ban held into the 1960s, when a few women were allowed to become corporate pilots. Most European airlines, however, did not relax their restrictions until the late 1970s, following the U.S. lead.

The glamour associated with the female flier masks the difficulty these women faced in practice. When it was not opposition from their families (who considered their dream akin to prostitution), opposition from aviation professionals themselves proved substantial. Like their counterparts in the United States, female fliers in Europe often faced sabotage and difficulty in finding sponsors. Because of the pressure they experienced, several tried even more dangerous records (one French aviatrix stayed up in the air for 30 hours and would splash cologne into her eyes so that the burning sensation would keep her awake). Two at least ended up committing suicide when their planes had technical problems, and most eventually left the field in disgust, realizing they had been admired as acrobats but were not really taken seriously as pilots.

The one aviation figure everybody agreed upon was Charles Lindbergh. His successful crossing of the Atlantic alone in May 1927 created an instant sensation across Europe. Boyish looks aside, his quiet demeanor clashed with stereotypical visions of Americans derived from World War I, and his solo success also contrasted markedly with those who failed ahead of him, mostly in teams of two or three. The American way fascinated, and this feat of aviation success called for emulation. There followed dozens of attempts over the following three years, most of which resulted in failure. Any who succeeded, however, could count on a hero's welcome in New York.

Other forms of heroic adulation also appeared in the form of groups of pilots. Mail carriers (which was Lindbergh's own earlier profession) became part of an elite that symbolized strength and determination. Their exploits would be recounted in schools, for their work was deemed essential to the maintenance of proper control over the colonies.

Most of the technological progress from the 1930s until after World War II came from the United States, where several landmark models such as the Douglas DC-3, DC-4, and Lockheed Constellation were developed. The first pressurized machines, allowing airplanes to fly above 10,000 feet and therefore reduce turbulence, appeared before World War II, but it was progress made with bombers (in particular the Boeing B-29) that allowed for the appearance of pressurization on passenger planes.

The post–World War II era saw new progress in the service given to passengers. With some exceptions, European airlines played one-upmanship with American carriers, especially Pan Am and TWA, though some introduced new luxury products in the form of special first-class service food to certain destinations. The international technical standards for flying were set to American ones because these had made the most progress. Consequently, European pilots, all trained in the metric system, had to relearn the old Anglo-Saxon measurements of distance and quantities (miles and nautical miles), all the while using the metric system, too, which is standard in engineering. The European airlines' preference also went towards flying with American equipment. There was a surplus of short-range aircraft available, and American manufacturers had mastered long-range flying without facing the destructions of World War II. Also, the competition from U.S. carriers was such that early forms of price warfare began to emerge in what was then a regulated market. Most airlines flying internationally were expected to be members of the International Air Transport Association (IATA), a private cartel of sorts that determined price range and offerings on all routes. It is with the approval of IATA that economy class was first introduced in Europe over the 1952–1954 period. This encouraged considerably more travel, and for the first time in 1958, more people crossed the Atlantic by plane than by ship.

The year 1958 also marked the beginning of the transatlantic jet age. Prior to that, the British manufacturer de Havilland had flown a model known as the Comet, which revolutionized air travel by carrying passengers twice as fast as existing propeller aircraft. Technical problems, however, caused two crashes and required the design be redrawn. In France, another jet design, the Caravelle, was built and flown in 1955. It was intended for short to medium routes and proved moderately successful. But the major production of airliners still came from the United States, with intercontinental Boeing 707s and Douglas DC-8 models. For shorter ranges, European airlines split their orders between DC-9s and Boeing 727s and 737s.

It is in the early 1960s that a new concept appeared on the drawing tables of engineers: the supersonic transport. The idea behind was that since early jets had doubled the speed of transports, it was only logical that the next step involve breaking the speed of sound with a civilian machine. The costs, however, were substantial, and only through cooperation might a successful project be achieved. This is what prompted the British and French governments to establish a collaborative program for the Concorde project.

Concorde took almost 10 years to fly and another 5 to enter service. Though considered a masterful technological achievement, it was an economic failure. It turned out that the future would be in bigger "wide body" transports such as the Boeing 747, not ultrafast jetliners. Other factors played a role, including the novelty of the technology itself. One engineer recalled years later how a British passenger had commented sulkily that there was no big difference between flying a Concorde and any other subsonic jetliner. The engineer replied, "that was the hard part to build" (Hooker 155). Indeed, the machine flew at 60,000 feet (most airliners fly at half that altitude), so high that the curvature of the earth is visible. Yet, at that height and speed, the pressure and temperature differences account for the aircraft actually lengthening by 10 inches before returning to its original size on the ground. Ideally, no windows would have been put in, but this contradicted the notion of proper passenger comfort.

Concorde arrived at the worst time imaginable. First flown in 1969, it reached production stage by 1973, at which point the oil shock made a mess of its economics. Simply put, the supersonic carried four times fewer passengers than a Boeing 747, but it used four times more fuel. Though it was no noisier at large airports than run-of-the mill jetliners, its climb to cruising altitude as well the sonic boom made it undesirable even to nonecologically minded people living in its path. It thus came to pass that only national carriers Air France and British Airways acquired it (all other airlines cancelled their options) and even got the last two for a symbolic franc in the late 1970s.

Once cast into a luxury bird, however, Concorde made a huge impression, akin in many ways to the luxury liners a half century earlier. The experience, however, was compressed into the 3.5 hours an Atlantic crossing took, instead of the usual 7 or 8. On its home turf, the fruit of international cooperation became a patriotic-themed machine, opening and closing air shows, transporting presidents or royal family members, and inspiring youth to work in aviation. The crash of a French machine in summer 2000 ended the dream. Though it returned to service after technical modifications, the combination of the September 11 terrorist attacks and international tensions combined to end the aircraft's service in 2003.

Nowadays, to Europeans, the major symbol of aviation tends to be the Airbus consortium. Conceived in the late 1960s as a means to help

Back to the future: A British Concorde is pushed back for its flight to New York City. The luxury speed service ended in 2003, but the symbolism associated with the machine remains strong in European minds. Source: Photograph by the author.

European aerospace gain some independence from the United States, the French-based international conglomerate produced a series of wide-bodied aircraft beginning in the 1970s. Initially selling poorly, the eventual success of later models (notably the A 320 and A 330 models) prompted some demonstrations of pride in European media. However, the lack of distinctness among most commercial airplanes produced worldwide (one would be hard pressed to spot a Boeing product from an Airbus one) does not elicit popular enthusiasm anymore. Flight has become routine, partly as a result of deregulation.

Europeans loved aviation early on, but most could not experience its benefits until the 1970s. By then, airlines began to offer grey market fares, whereby heavily discounted tickets were sold through travel agencies and social centers. Smaller charter companies did appear on the scene but had to fight so-called flag carriers to be allowed to service even destinations the flag carriers did not serve. The U.S. deregulation act of 1978 started the ball rolling toward a liberalization of commercial aviation. Airline mogul Freddie Laker, who ran a namesake charter airline, introduced a revolutionary Skytrain service in 1979 between London and the United States. The no-frills approach included bringing one's own lunch and entertainment, as all movies and music on board were removed to save money.

Some crew were known to initiate joke contests, and most passengers went along with it. Unfortunately, members of the IATA cartel matched Laker's fares on the grey market and sued him in court. The impact of the second oil crash finished off his grand design.

It was a decade before other airlines, mostly charter services, modified their offerings into scheduled flights. As they did so, however, they drove some established carriers from strong markets and became favorites of budget-conscious consumers. EasyJet, for example, chased out Air France and Swissair services between Geneva and Nice (French Riviera). Whereas a ticket on the latter cost as much as $400 for a one-hour flight, new fares reached as low as $20. In saturated airport areas, budget airlines also sought to use old military bases and third-level airports to cut costs further. Thus, passengers flying the low-cost Ryanair out of Frankfurt, Germany, found themselves hopping on a 90-minute bus ride to Ham airfield, whereas the main Frankfurt airport was but 5 minutes to downtown via express train. Most did not seem to mind, however. Unlike their parents and grandparents who had known trains and buses at most, many now experienced flight. Unglamorous though it is today, the element of speed on the way to a vacation spot more than makes up for the seeming gloom of airports.

Fifty years separate this photo from the following photo. The first depicts the Finnish airline's inaugural flight between Helsinki and London using an American-built Convair. Passengers, mostly businessmen, are properly dressed for a workday. Source: Courtesy Finnair.

This picture shows two of EasyJet's planes awaiting their load of passengers at Amsterdam. The routine of air travel has made such airlines popular in Europe, where the extremely low fares prompt consumers to accept the inconvenience associated with steerage treatment, including crowded, hangar-like terminal areas. Source: Astra, Geneva, Switzerland; reproduced by permission.

Airports in Europe differ notably from their American counterparts in that for decades, many of the former were considered places to relax or entertain oneself, in the manner of a market square or a shopping mall. The architecture of airports proves it: Until the 1970s, most European airports sported at least one wide observation deck with a restaurant where families would come on days off and watch air traffic. Tragically, a series of terrorist attacks at airports in the 1970s saw terrorists throw grenades from observation decks at specific aircraft. Many decks thus disappeared. Some major hubs still offer that option nowadays, conscious that it is also a good public relations tool, as the taxpayers' money had often gone into building airport infrastructures. Amsterdam and also Frankfurt-Main and Zurich are but a few examples of this. The public relations dimension has gained in importance due to ecological challenges to airport expansions.

Until the 1960s, airports were often planned and built with very little attention to the immediate surroundings. The noise disturbance associated with jets, however, prompted many communities to impose restrictions on flights at night, but it was the ecological movements of the 1970s that proved the greatest challenge. Like their counterparts in Japan (Tokyo-Narita airport) and the United States (Boston-Logan), opponents focused especially on the addition of runways. Unlike a terminal, a runway runs at least two miles in length, to which one must add safety zones, approach lights, and beacons, to name but a few. One of the bitterest fights involved a Frankfurt airport in the early 1980s. There, local residents rose against the planned establishment of a new runway perpendicular to the existing installations: "18 West" as it became known would cut a huge swath of cement into pristine forests. The opponents, many of them women but also church goers from villages scheduled for destruction, braved subzero temperatures, invaded sealed up areas of the airport, and used legal means to try and stop construction. Though they failed, their experience created tensions that last into the present day over the growth of the giant airport. Nowadays, passengers use Frankfurt airport very heavily for business but also for leisure travel. Leaving the airport using any of the runways they can see during their flight's climb the impact 18 West had on the surrounding forest. It represents symbolically the costs associated with the ease of flying that entered European culture starting in the 1980s.

CONCLUSION

Sea travel had long been routine, but it disappeared in favor of air travel. What remains of excursions on water is often found in the realm of nostalgia and tourist trips, such as boat rides along Europe's major river axes. As for longer distance sea faring, with the exception of a few ferry services (such as the Channel ferries or those between islands), it has been replaced with speedier means of travel. Automobiles and trains use tunnels or bridges, and the plane has made the sea faring experience obsolete, except in niche markets associated mostly with leisure.

As for aviation, its recent appearance in all its forms revolved first around the notion of spectacle. Much in the same manner one watches a lion tamer, one considered pilots with similar admiration and puzzlement. Though aviation has become routine, the air show itself remains, almost a century later, the embodiment of the notion of spectacle. Second, aviation had to convince. It could not simply find a use immediately, and thus confirms the reversed credo that historian Melvin Kranzberg argued, namely, that invention is in fact the mother of necessity. Once it had convinced Europeans that air travel was good, though, commercial aviation especially became routine. Airports alone and perhaps the experience aboard different airlines indicate a difference of experience.

6

COMMUNICATING

Hand in hand with physical communication, Europe experienced the gamut of communication tools. Some were meant to share information instantly, like the telegraph; others to store it and support other enterprises, such as the typewriter. Several came to entertain. All were adopted in various fashions and not always with the expected level of success.

THE TELEGRAPH

The telegraph represents a special case of the applications of science to technological invention. The result of several scientists' work across borders, the telegraphic concept of transmitting sound using electricity gained fame once a new means, Morse code, was introduced. Samuel Morse in the United States tested his invention in 1843. The stupendous implication of this means of communication became quickly clear to governments in Europe. Until then, many had relied on Claude Chappe's signaling system, an optical telegraph first devised during the French Revolution, which allowed distance communication between raised stations (on hills or towers). Though a substantial jump in long-distance communication, most Europeans did not register any impact on their lives other than seeing oddly shaped towers in some regions (Standage 21). It would be otherwise with the telegraph.

In the case of Morse's invention, when combined with the perfecting of long-distance electrical signals, the telegraph came to be considered a public utility that businesses as well as individuals embraced. The latter came to appreciate its speed and, despite its high cost, often used it to communicate urgent news over distances the phone could not cover, or would prove too costly. Culturally, traditions also were developed. Condolence messages for funerals were expected and considered a sign of respect for the departed. More happily, telegrams at weddings came to be the norm and were traditionally collected by the best man for public reading between dances. The tradition disappeared in the 1990s when post offices stopped accepting telegrams due to the competition from the Internet.

Businesses, too, took to relying on the telegraph and went so far as to pay for private lines that would include special daily updates on market trends. To this day, established companies keep a variation of the telegraph handy: the telex, or telescriptor. Its technology, which combines the principles of the telegraph and the telephone, involves the printing out of messages on electric typewriters. A standard of governments and the military (the so-called red phone between Washington and Moscow is in fact a telescriptor), it lost favor to the fax and the Internet. It remains in use among some companies because messages sent using the telex landlines are considered legally binding.

The telex also helped spread the news business; newspapers would eventually subscribe to newswires, and this would ensure the spread of the tabloid press in the late nineteenth century. This was not the only factor, however. New development in typographic presses made possible the jump in periodical printing and sales in the second half of the nineteenth century. Rotating presses, which allowed the feeding of huge rolls of paper into a printing unit, cut on the need to measure the paper and place sheets under the printing press. Coupled with the industrial diffusion of ink and the advent of mechanical type setting, the speed of publication increased substantially. Consequently, newspapers also began to sell advertising to increase their revenue and could promise swift reproductions of texts handed in. Among the advertisers, movie producers and theaters came to make heavy use of newspaper ads to encourage consumption of the moving image.

By the 1920s, movie theaters began competing directly with theaters for entertainment. An invention of the late nineteenth century in both the United States (Thomas Edison) and the Lumière brothers, movies and filmed news events remained silent, with narrations read over the screen occasionally (or played on a recorder). Major production units (such as Gaumont and Pathé in France) also tested new methods of entertainment. Emile Courtet, aka Emile Cohl, completed one of the first cartoons in 1909, *The Happy Microbes,* a humoristic snapshot of humanized bacteria.

The biggest success, however, became filmed news, where Pathé-Journal (distributed in the United States as *Pathé-News*) became a big draw as a prelude to the main entertainment. As for movies, after World War I, some 70 percent of all productions distributed in Europe were American ones, where the public discovered the likes of Charlie Chaplin, Harold Lloyd, and Buster Keaton.

European production, however, did include major figures, too. Some took on the matter of technology. In Germany, Fritz Lang filmed *Metropolis*, about the extreme mechanization of a future—yet timeless—city. His subsequent movie *Frau im Mond*, though launched in the midst of a space craze in interwar Germany, failed to register the same impact his earlier dystopia did.

RADIO

The turn of the century marked the slow advent of the radio. Though the transmission of information was first suggested by Nikola Tesla and later turned into a profitable operation by Guglielmo Marconi by 1907, most public programs did not appear until after World War I. On the other hand, amateur radio operators began showing interest in the system in the 1890s. Marconi disliked that his system could be heard by all, unlike the telephone, which offered a measure of privacy. Still, early aficionados ignored the issue: Meeting complete unknowns on the airwaves was exciting, and many amateurs thus learned Morse code to communicate. In fact, it was during World War I that amateurs showed how it was possible to broadcast voice effectively.

Right after the war, European governments began legislating the airwaves and removed chat operators from easily accessible airwaves, relegating them to less used ones in favor of public broadcasts. Often at first, these were presented as publicity stunts to encourage people to buy receivers. In France, the first radio shows were done in 1921 from the top of the Eiffel tower and were followed the next year by a private channel, Radiola, which marketed receivers, too. The challenge for early radio consumers was to be able to manipulate their receivers easily. Soon, various brands introduced the single tuning button, or pretuned station receivers. Costly, the receivers still required the use of heavy batteries. Soon, however, the first electric-powered receivers appeared in stores.

Initial reactions to radios in the home were, like to many other technologies, a mix of awe and distrust. Early radios were very much an assemblage of parts that would look better in a garage or a proverbial mad scientist's lab. The goal of radio manufacturers, however, was to convince consumers that such contraptions would become a new source of entertainment. Consequently, the radio receivers could either be "dressed up" as furniture or designed as new items for display in the home. Many of the early models were indeed enclosed into classically sculpted cabinets.

Radio was a novelty, yet, one everyone wished for. In the case of Great Britain, it was actually the first electrical equipment to be owned on a mass scale. This is all the more remarkable in that throughout Europe, ownership of wireless (and later television sets) required the payment of a license fee that would cover broadcasts. Yet, in 1922, when broadcasts began, there were already 36,000 licenses issued (these were bought at the same time as the receiver). By World War II, there would be 6.5 million. When facing a choice, people of limited means preferred to invest in a mass-produced radio receiver than in any other electrical apparatus (Forty 200–202).

A new culture of the wireless thus developed, one in which the radio became a social gathering place to hear music but also news and commentary. British citizens took to listening to regular radio addresses from King George V. In 1936, they also listened in as the radio announcer stated, "His majesty's life is drawing to a close," followed by classical music. Two years later, millions of Europeans gathered around radio sets to hear the results of the Munich Conference to find out whether there would be war that year or not (Prost 133). The same radio sets broadcast Marshall Pétain's call for an armistice in France in June 1940, and it was BBC broadcasts picked up discreetly at night during World War II that gave hope to many more on the continent.

The case of Nazi Germany and its national radio sets, known as *Volksempfänger* (literally "people's receiver") is notable as an attempt to draw as many consumers into the fold of a technology. Indeed, whereas in 1933 only 25 percent of German households had a radio, almost 75 percent owned one in 1941. Though seemingly impressive, the number nonetheless reflects limited success. One reason for it was that several privately produced radios remained available for sale despite the Nazi regime's stress on the need for a proper *Volksempfänger*. The latter was supposed to be cheaper, but the German population associated radios with middle-class status, which meant many workers chose to invest their money elsewhere or to listen to the radio in community settings. Another reason was the licensing fee. Every radio required the payment of a tax, which was used to finance the propaganda ministry's programs. Hence, it comes as no surprise that, while many households eventually acquired a radio, they did so at a slower pace than in democracies like Denmark and Sweden in absolute numbers, or Norway and France, both of which achieved higher growth rates with less state investment (König 84).

Sports were also broadcast and became the first grounds for spontaneous celebration (and commiserating) away from the site of the actual competition. Boxing matches were among the favorites of listeners, but so were team sports. The 1954 world soccer finals, for example, saw Germany as the underdog beating Hungary. The entire German commentary, broadcast live in homes, pubs, and radio stores (there was no television coverage), captivated listeners nationwide. Herbert Zimmermann announced from the stadium in Bern that "that which we had feared has come to pass," referring to the early 2–0 lead by the Hungarians, only to then bring Germans into hysterics when he announced less than 80 minutes later,

"The game is over! Germany is World Champion! Beats Hungary 3 goals to 2 in the Final in Bern!" (Courtney)

Away from the excitement of sports, yet just as entertaining, were plays. The first were broadcast in 1924 in England, France, and Germany, and soon prestigious actors and composers were solicited to produce made-for-radio plays. Such entertainment paralleled classical theater of course, but film, too. It was also deemed more democratic by some, as anyone with access to a radio receiver could enjoy a play without having to dress up and go out. Shakespeare's *Macbeth* was adapted to the airwaves in 1927. The seeming realism drew listeners and encouraged a different kind of escapism. Actors slammed minidoors made to sound like real ones, musicians played animal sounds, and a whole new genre was born. Just as Orson Welles's *War of the Worlds* broadcast in 1938 convinced some American audiences that Martians were invading, British listeners felt trapped in an underground mine when listening to Richard Hughes' *Danger* (Forty 202). On the continent, several playwrights gained fame through the broadcasting of plays written on commission. German Bertolt Brecht eventually staged some of these *Hörspiele* ("listening plays") in theaters. Wolfgang Borchert's *Draussen vor der Tür* [Outside by the door] about several characters surviving in Germany after World War II even went from being a radio play to becoming both theater and movie classics. In Switzerland, much of Friedrich Dürrenmatt's work came to be seen as interchangeable between sound and stage. Dürrenmatt also took on technology as a theme (see the discussion of the automobile). Of course, not all plays became classics. Audiences enjoyed as much soap operas and mystery plays as they did classics. In Germany alone, in over eight decades that radio plays were broadcast in the twentieth century, some 50,000 plays were heard. Some 200 might be considered classics (Thalheim). In many ways, though, they prefigured the audiobooks common in the Western world nowadays. Experiencing a story around the radio set was a special treat, but so was the fact that one could do it in chosen company. The gathering around the radio bore a definite social importance, and the culture of neighborliness developed into a full-fledged tradition that involved a light meal, or simply an after-dinner activity that was cheaper than going out. It also meant a willingness to share one's private space for the sake of a public sphere event. This is partly why receivers were often ornate, as they became not only part of the furniture, but reflected the taste and financial means of their owner.

As far as the receiver itself went, its design depended on competition among manufacturers. While the cabinet version was favored as blending into existing furniture, many actually preferred a new design that identified the radio as such: to own one was to be modern. This notion of modernity helps explain why, despite the fact that there had been few improvements to radio since the 1930s, acquiring such an item for one's living room was still akin to gaining new status. The expense came not

A middle-class northern Italian family in the 1940s poses in their apartment. The radio with a record player has become an ubiquitous part of the living area's accessories. Source: Private photograph, author's collection.

only from what was inside the box, but from the box itself. Until the 1930s, the only material used was heavy thick wood (to withstand the weight and heat of the elements), usually carved and varnished by hand. The advent of a new substance, bakelite, changed that.

Patented in the United States in 1906 and spread worldwide, Leo Baekeland's invention had the quality of sturdiness and was a good insulator. It replaced wood in many models, keeping the former reserved for high-end models. Variations of bakelite such as philite and arbolite (both used by the Dutch Phillips corporation for its radios) also appeared on the market. The colors available, though, were mostly of darker tones. With the arrival of new synthetic materials, new colors, but also cheaper production costs, encouraged the switch. Bakelite radios (also produced in new colors) did become part of the furniture environment in households but disappeared from production by the 1950s. Plastics were not the only reason for the switch.

Miniaturization changed the radio. Transistors replaced vacuum tubes, thereby making radios lighter and mobile. These were still fairly big but could actually be moved around the house. The cost of such radios, however, remained high. A blue-collar worker could easily spend one month of her salary in the late 1950s to buy one! Yet, the perceived freedom it afforded in moving about was all that it took to convince many to make the investment. Throughout, radio manufacturers emphasized a technical look for machines that were in fact very easy to use. Europe thus came to experience Japanese technological advances through radio first.

In 1955, the first Japanese-built transistor radio appeared on the market, followed two years later by the first truly portable machine. The Japanese actually inaugurated the practice of baptizing their models with numbers, which surprised consumers at first but was quickly accepted as a more modern way of buying a technical item. Local manufacturers opted for a mix between names and numbers, such as the British Pam 710. The evolution of names, though, followed the Japanese marketing trends. The first Italian transistor radio bore the name Zephir. Its main successor bore the far less attractive name TS 502. The latter, though, following the Italian industrial marketing trend, had a box drawn by Marco Zanusso, a leading designer. Once again, radio technology required a proper presentation to attract consumers as did its new complement, the television.

Independently of the clarity of reception (which depends on the power of the receiver and the quality of the elements used, but also the area where the receiver is), radios' diminished size made them attractive to take along. Some began to appear on beaches and at other vacation sites. For the more fortunate, the optional radio equipment was added into new cars. The small size of the radios made them ideal means of entertainment and socialization for youth. Army conscripts in particular came to prize such items, as they provided an important link to the civilian world.

In the case of French draftees, small receivers proved key in explaining why the conscript part of the French army generally refused to support a coup in 1961. To protest the French government's decision to give up the French colony of Algeria, several senior army officers attempted to stage a coup but required the logistics of the army divisions under them to complete the plan. Because draftees serving in Algeria had access to civilian radio, they became aware of certain information that was not available to them otherwise, including the fact that only part of the army was intending to strike at the civilian government. The uprising failed. Though radio was not the sole cause of that, it played a key role in its undoing (Prost 135).

By the late 1960s and into today, radio listening became more individualized, thereby following trends set in the United States. The eventual liberalization of radio airwaves (states began giving up their monopoly in the 1970s and 1980s), partly in response to popular preference for private programming, also changed the landscape. Instead of having simply access to public-funded stations with no advertising, younger generations favored commercial outlets that also broadcast music that was not found on public stations. The consumer trends favored American and British rock 'n' roll but also local artists who mimicked Elvis Presley, the Beatles, and other English-speaking entertainers. Radios in cars gained highly in popularity, especially in Italy. Paradoxically, this expression of individualism also paved the way for canceling out radio as a social instrument around which everyone gathered. The dual introduction of the Walkman tape player and receiver in the late 1970s, together with the advent of so-called boom boxes that also incorporated receivers, completed the process. The first kind required small headphones and introduced new dangers, as listeners on the street isolated themselves and lacked enough hearing to heed dangers on the road. The banning of headphones while driving did help reduce risks. Boom boxes, aka portable stereos, came to be seen as a public nuisance, and cities throughout Europe invoked laws against noise and also public loitering to check the other kind of socialization that came with playing one's stereo in public. The overall result was that, from a private tool in the home that introduced new notions of neighborliness, the radio had evolved into one used in the public arena, yet, one that did not imply automatically human socialization.

A further evolution in the late 1990s involved new competition and digitalization of programs. Though competing programs had already appeared in the 1980s, forcing public stations to shift their content in order to remain current, it was television that brought about new broadcast conditions. Until then, Europeans traditionally listened to morning radio during breakfast or during car commutes. The introduction of morning TV programs in private, then public, TV stations forced a reexamination of radio content. That is still ongoing.

The other new challenge came from digital programming. In some ways, this constituted an extension of cable radio, a concept several decades old and used with dedicated receivers, notably in Switzerland. There, though, the receiver was the computer.

It is worth mentioning one particular aspect of radio operating, namely Citizen Band (CB), a hobby that dominated in the United States but also found favor in Europe. Though identified with amateur operators, the culture of CB involved many rules, traditions, and advanced technology, at least to the degree allowed by law. Indeed, as European nations took monopoly control of their airwaves in the 1920s, they left only small swaths of airwaves to CB. After World War II, though, the law changed and allowed for voice rather than signal communication alone among amateurs. Their numbers rarely exceeded several thousand in each European nation, but a kind of heroic myth grew around the CB community. Two radio amateurs in Italy, for example, tracked the launch of the Soviet space probe Lunik IV in March 1963, beating the official press communiqué by 24 hours. Aside from news coups of such kinds, it was the language used to communicate that made CB practice an odd subculture. Every radio operator had a code name (usually registered with the radio authority of each country). As for the vocabulary in use, it was reminiscent and sometimes directly taken from professional radio. But what did they talk about? The content of messages was strictly legislated. No national news especially was to be discussed, as it was considered competition for national radio programs. There is an irony here, for a new competition under state monopoly appeared in stores after World War II: the television.

THE IMAGE BOX

Whereas transmitting voice sound had become possible by World War I, sending images was another issue. In effect, the optical illusion that gave the human eye the impression of mobility in movies (at least 16 frames per second) had to be replicated through the airwaves. Scotsman John Logie Baird came up with a system that scanned the subject of transmission through a series of tiny holes and transmitted the scans. Though complex, the system worked but required the subject not move too fast (Buchanan 164). This challenge prevented the system from being adopted in Great Britain (the receiving apparatus would have been just as difficult to operate). Aside from Baird's contraption, most attempts at building machines capable of such speed failed: Electronics were required. The Russian-American Vladimir Zvorikin built an ionoscope in 1929, which looked like a tube that could decompose and reconstruct the image at the necessary speed. A key element to the design of TV had finally come about. Further experiments in the United States and Europe allowed for the first experimental broadcasts within a few years.

The first full television broadcast in Europe happened at the Berlin Olympics in 1936. However, the limited number of receivers meant that the experience had little impact on the German population: Radio remained the main source of information and entertainment. The mass-production of TV receivers in Germany was planned, with some 10,000 units of the first standard model, FE-1, to appear on the market in 1939. In fact, 50 left the factory before the tanks started rolling. Europe in fact followed far behind the United States in the establishment of programming and the availability of television receivers.

Television's entrance into the European households was slower. As late as 1960, acquiring a TV set in French provincial cities was as much a source of pride as of embarrassment. One survey showed inhabitants asking the electrician in charge of setting up the set to keep quiet about it and let neighbors notice the roof antenna on their own: An intruder, TV, was entering the community. There would be other means of entertainment besides visitors in the flesh. At the urban level, younger generations had welcomed the TV set, and many had watched the first live transmission of a political meeting, in that case the handing over of the French presidency from Vincent Auriol to René Coty in January 1954. That year, though, barely 1 percent of all households had a TV. By 1960, it reached 15 percent, then 67 percent by 1970. Socially, the family remained nuclear but crowded around the single TV set. By the 1990s, 40 percent of French households had at least two TVs, thus introducing a new social dimension to technology: watching TV alone. The acceleration of miniaturization is further shifting the trend (iPods).

The high cost of the image boxes, as some called them, meant that to own one was as much a matter of status as of taste. The British satirical magazine *Punch* depicted the dilemma in a cartoon where a woman asked an electronics salesman for a price quote on the installation of a roof antenna without the TV set, one that would look just like her neighbor's. Yet, the television market was deemed to have such potential that technological battles were fought away from the public eye resulting in three major systems coexisting to this very day: PAL, Secam, and NTSC (the U.S. format). By the 1980s, when VHS-format video recorders gained in popularity thanks to lowered costs, consumers experienced the frustration of being unable to play American videos on European machines, and vice versa. Two decades later, a similar battle was fought with the creation of DVD regions to check the illegal imports of DVDs from one cheaper market into another.

The television set, when it did become part of the household, initiated social changes comparable to those that affected American households. First was the development of a new kind of neighborliness. Initially in existence when radio appeared in the interwar years, the practice of having the neighbors over to watch TV became common in several societies, notably Germany.

The events that drew people together varied. Some, such as major concerts, echoed radio tradition. In the realm of sports, however, the introduction of *Eurovision,* a television consortium to allow broadcasting of major events, changed everything. The first such event came on June 2, 1953, the crowning of Queen Elisabeth II of England. The limited number of television sets privately owned did not reduce interest: Bars, hotels, even newspapers placed sets in their windows or high stands so all could follow the event. Later, in 1960, the Rome Olympics were the first to be seen live by spectators outside the stadium. State funerals, royal weddings, and major allocutions were also important. In July 1969, neighbors woke each other up to watch Neil Armstrong become the first man to walk on the moon. All these events, framed within newly designed talk shows, news segments, and entertainment stories, turned the image box into a new form of entertainment, much the way the newspaper and radio had been.

Whereas the radio continued to evolve successfully into a central element of entertainment despite TV's presence, newspapers did not fare so well. The European press in the post–World War II era remained centralized around major newspapers in each country, with occasional local publications that sought to emulate the big papers. Tabloids became rapidly successful for addressing everyday concerns of readers and were able to negotiate the TV challenge in the United Kingdom and much of Western and Central Europe, but in France, where a tabloid press failed to develop, the established press lost both readers and journalistic talent to television.

Other factors played a role, but the small screen was instrumental in accelerating the decline of steady readership.

Second, like in the United States, a new member of the family was present at the dinner table. The broadcast times for the news reflected this habit (7:30 p.m. for Switzerland; 8:00 p.m. for France and Germany). Many families rejected this but often found that, on occasion, the practice helped relax social tensions. The dreaded Sunday dinner with relatives could be entertaining for adults, but children often shuddered at the hours spent at the table. The TV set resolved the tension, even though broadcasts were few, and TV was black and white (color appeared in the late 1960s) and might involve as much entertainment as it did politics.

Third, though the use of television for political purposes had been presaged in the 1930s and proven by the early 1960s, political debates were few. Legislative discussions eventually appeared, as did major addresses by national leaders, but the notion of debating candidates, for example, was deemed Kennedyesque, an ambivalent judgment that reflected both worry and admiration for the American president whose TV performance had charmed Americans and Europeans alike.

In 1965, for example, France elected a President for the first time by direct universal suffrage (a change in the country's democratic tradition). However, voters favored the incumbent even though the latter seemed least approachable on screen. The access to politics and news on a small screen began to affect movie theaters, which had traditionally served as news outlets since the 1930s. Newsreels disappeared completely by the late 1960s, and many movie theaters had to close. Those that survived required the installation of bigger screens, more comfortable seats, and the introduction (based on the American experience) of multiplex theaters (Buchanan 173).

The concern over youth exposure to television appeared early on and paralleled the concern over corruption by youth from comic strips in the 1950s. This tension over visual culture, of which television was a prime component, reflected as much a malaise over the shifting relationships between adults and children as it did concern over technology. Perhaps best reflecting this state of mind, many European TV manufacturers offered television sets with a lock on the control pad. The feature disappeared on models with remote control, but costs associated with the top range of television sets meant that the lock feature was found in many households well into the 1970s.

However, the limited number of channels and of air time as well as state control of most channels in European countries meant little evidence of youth corruption actually surfaced. Governments maintained monopolies on broadcasts and applied different standards in the guaranteed liberty of expression from those that applied to the printed word. A whole series of movies were banned until the late 1970s and early 1980s as too controversial for TV audiences, even though they could be seen on movie screens. For example, the BBC's *The War Game,* a docu-drama on nuclear war, was banned from the airwaves in 1966 for almost twenty years. Marcel Ophüls's *The Sorrow and the Pity,* a 1969 documentary on French collaboration, faced a similar fate in France until 1982. The relaxation of such seemingly unwarranted bans reflected not only a shift in social and political outlooks but also one in the spread of media outlets.

In the 1980s, the increased deregulation of broadcast rights along with the appearance of affordable cable and videotape recorders brought about changes in youth culture. Sociologists began identifying the emergence of separate youth groups mostly among teenagers who used the new flexibility of the medium to develop further their interest (Hutchby and Ellis 34). In this case, though, television became a catalyst for sociability rather than a hinderer thereof. As in the United States, the development of fandom for a certain series became common practice. However, television also came to share the scene of electronic entertainment with video games and computers.

COMPUTERS

As two sociologists put it, "information and communication technologies (ICT) are set to wreak widespread social, cultural economic and political change in the twenty-first century" (Hutchby and Ellis 58). Consequently, computers in particular came to assume status symbol much in the same manner television sets or even washing machines had four decades earlier. Although computing technology by the 1990s remained far from mature, thus rendering new machines obsolete in a matter of years, many families felt pressured to acquire a machine so that their children would be techno-savvy. This attitude reflected a lack of understanding of what ICT could and could not do. Yet, acquisition of a computer also sowed the seeds of conflict, whereby belief in using the machine for educational purposes clashed with youth perception of a computer as a new platform for video games.

The advent of computing in everyday life has to be considered from two distinct angles. The first, the advent of computing machines for governmental purposes, the second the introduction of microcomputing beginning in the late 1970s.

In the realm of the first, statistical analyses and projections such as population growth, market studies, and urban planning all contributed to the computer leaving the scientific realm for administrative applications. However, the conversion to computing was slow and was dominated by machines imported from the United States. By the 1960s, IBM came to dominate the European market with its 360 series of computers, easier to use and nominally using the same program no matter what the model (a radical improvement in an era when programmers had to be retrained for each new model purchased).

Several reasons help explain the slow growth of computing, but it is generally narrowed down to three: The costs associated with developing computers meant that few countries with limited markets could afford to seek companies willing to invest in a new technology. Second, the technological gap was substantial by the time indigenous European companies caught on. Third, the lack of public understanding about what computers did. To many, a computer looked like a series of boxes doing intricate calculation, displaying dials, and little else. The programmers were considered just as mysterious, and little of substantial value could be shown for all the operations performed.

With IBM's entry into the European market and the progressive acquisition of small (non-PC) computers by private companies, a greater awareness of computers came about. No longer was a computer simply about printing one's check on payday or calculating an algorithm. Instead, information storage and management became the keywords of successful industry. The oil shock of 1973 accelerated this awareness, for the computing industry, still much smaller than its American counterparts, was among the few to still turn a profit during the ensuing economic crisis.

In this context, the advent of microcomputing was remarkable. As soon as the first microprocessor, the Intel 8008, was invented in the United States, a few European engineers imagined an unanticipated use for this device: The microprocessor could become the central unit of a computer. The first microcomputers (the term was coined for the patent application) were produced and sold in 1973 in the Paris region. Several small businesses followed suit, developing various models and incorporating the new generation chips and the market needs of clients who might afford such machines. Generally though, the costs of each model remained extremely high for very limited returns.

Among the few exceptions, the 1981 appearance of the British company Acorn's Proton, also known as BBC, stands as a notable success. What made it successful despite its price (as much as US$1,000) was the fact that the BBC would broadcast computing classes that used the same machine and that schools eventually adopted it for basic programming instruction. Furthermore, a series of programming modules were also made available so that once bought, there was a genuine option to use the machine rather than let it sit and collect dust.

When support and instruction were not available, enthusiasts were soon able to rely on low coast. The defunct brand Tangerine commercialized ORIC 1 in France and England for about $300 a machine. Its low cost was due in part to its very cheap design (most users complained about a dreadful keyboard) and the fact that data was to be saved on a tape recorder at a time when floppy disks were becoming available.

Paradoxically, just as some parents eagerly sought to acquire high-priced computers, teachers surveyed in Great Britain appeared less enthusiastic about using terminals. Not only was the need for new training an issue, but there also was a fear of losing central status in favor of screens (Hutchby 64). Meanwhile, in France, government subsidies allowed the acquisition of many computers to schools, but the lack of teacher certification in these fields meant many such machines ended up in classroom cupboards. What eventually made the general public aware and more interested in computing were the appearance of computers with friendly graphics, such as the famed Apple 2 in the late 1970s, and the standardization of programming through IBM, which adopted the Microsoft MS-DOS standard in 1981.

A related concern began to appear at the level of consumers. In the 1980s, early studies of young computer users suggested a gender divide whereby girls were familiar with basic computer skills, but not comfortable with them, whereas boys spent more time playing computer games and thus gained familiarity with their machine's dynamics (Hutchby 67–69). The corollary, though, was one familiar to American society: the appearance of the so-called computer geek. Typecast as socially awkward and out of touch, the computer geek was in fact reflective of a certain unease with computers. The inability to understand programming prompted mass culture to project anxieties onto the ones who could actually figure out computers. Ambivalent feelings of disdain and admiration combined and echoed those encountered in the United States when dealing with techno-savvy individuals.

Technophobia relating to computers, however, became affirmed when various scandals about Internet child porn and other sexual encounters began to surface in the late 1990s. The 1996 World Congress against the Commercial Sexual Exploitation of Children in Stockholm identified child pornographic abuse of the Internet as a growing problem. As a direct result , in November 1999, the European Union initiated an Action Plan for Safe Use of the Internet. It has a comprehensive Web site and continues to attract new members from around Europe. To be effective, however, it also relies on individual action and asks for direct reports to take action against predators through its "Daphne" program (Playing Safe).

THE PHONE

The child of Alexander Graham Bell (at least in terms of the overarching goal, distance voice communication), the telephone rapidly gained followers and subscribers in Europe. Governments, initially willing to grant licenses to private company to operate city exchanges, soon saw the strategic and economic value of the system and moved by law to construct a national monopoly. In certain countries, it remains in effect to this day.

The technology initially relied on human contact beyond the caller and called: Operators, mostly male, were summoned when the phone was unhooked and proceeded to connect the correspondents. By the 1890s, and fully after 1900, telephone services were authorized to hire women, as

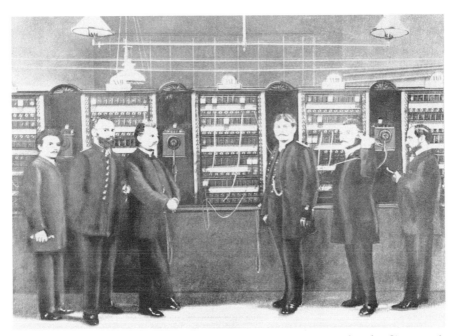

The first telephone exchange in Germany was built in 1881 by the Siemens & Halske Company, under contract to the government. Initially, men alone worked there, but within two decades, women generally were assigned to these functions. Source: Courtesy Siemens Pressebild.

their salary was lower and the task was considered manageable at their level. The sexism associated with this move was a double-edged sword, for, like the secretaries who replaced male clerks at the time, women who operated phone lines gained a certain financial independence, which they could keep until married (in most countries, such as Germany, women could work between ages 18 and 30 by which time they were expected to have joined in matrimony). Socially, they also controlled every phone communication, even though they were under male supervision. Callers came to expect women as operators, preferring them to male operators.

Calls invariably started in the same way. "What do you wish?" asked the operator when summoned. She would then try the link (a matter of plugging in a few connections or calling on a colleague across town to complete the link to another area). If the line was busy, she would report, "Already occupied; I will call you when it's free." Once the call had ended, the operator was expected to disconnect the link.

Callers did not sit idle, however. Nowadays, landlines rely on central-ized power (independent from the electrical grid) to carry the wave. Up until World War I, many consumers were expected to purchase and main-tain their own battery to feed their phone line. A small crank was con-nected to the phone to generate the "call current" that would start the battery and advise the operator a connection was needed. If the phone rang at the called and the latter answered, the operator announced "I am connecting" and stuck in a second plug.

All nations proceeded with this system, and some maintained it for decades. The modernization process was haphazard and slowed by a combination of financial limitations, monopoly control, as well as the two world wars. There also existed a fear of the unknown. The phone connection was carried out by a human being. Wrong numbers happened all the time, but there was someone there to help or correct the error. The switch to a machine prompted many to argue that this was prone to end in more errors in dialing. Advocates, however, were able to make their voice heard, and one summarized the advantages of automatization:

> Communications are quickly obtained, especially during moments of intense traffic. The phone line is available at all times of day and night. The automatic unnecessary waits no longer happen. Connection mistakes are impossible: only the dialer's finger can be blamed. Finally, the secrecy of the phone call is guaranteed. (Cramois)

The conversion remained slow. In Germany, for example, the first automatic connection was set up in 1908, but the last human-operated one did not stop until 1966. France initiated the first switch to automatization for 40,000 of the 200,000 Parisian phone owners in 1926, but in 1931, the first "slice" was not completely equipped. After World War II, comedian Fernand Reynaud, commenting on the fact that his country had kept up with international communications and could connect Paris to any capital in the world in a few seconds, also noted humorously that for someone to call Aniere (a few miles away from the French capital) could, on the other hand, take hours at the hands of local operators. French local and regional lines remained in this state of affairs for a very long time, which also explains the paralysis of the nation during the student riots of 1968, when telephone operators went on strike and blocked automatic connection centers and helped practically shut down the nation for a few days.

The situation for the French began to improve in the 1970s with the laying of new plans for a more efficient phone system, which was put in place in the 1980s, along with the introduction of an ancestor of the Internet, Minitel.

Enter the Cell Phone

Likely the fastest success in communications technology, the cell phone went from being viewed as an elitist tool for showing off to a standard service, deemed practical by households and essential to youth. The parallel to the American case is striking, but the spread of Handy (German), Telefonino (Italian), or simply mobile (France) was much faster in Europe than in the United States. One reason for this was the willingness of monopoly companies to grant operating licenses promptly. Consequently, it became more practical for households in remote areas to get linked in a matter of days rather than months. From the early to the late 1990s, increases of as much as 90 percent were registered in the number of cell phone owners in Europe. Though it had started as a convenient tool for business, replacing beepers as well as the need to find a phone booth, the low cost of a basic cell phone (though not its actual calls) convinced families to obtain one for use in emergencies. Further usage eventually came, and the new technologies that included text messaging became popular with youth. By the close of the century, the market for cell phones was considered to have reached saturation among most levels of the population: Only older citizens remained suspicious of its use. Sociologists and philosophers alike have detected a new manner of behavior in relation to the mobile phone similar in many ways to the concerns raised in the United States. These include worries about how much autonomy to give youth using text messaging but also how there is now a need to be reachable. The corollary, that anyone calling you can expect an answer rather than an answering machine, also raises issues of privacy and the right to time away from work (Goggin).

Phone and Terminal in One: The French Minitel

First experimented on in 1980 in the French region of Brittany, the Minitel system was originally intended to provide directory assistance and business information with the goal of eventually replacing phone books, deemed too expensive to print. Instead, it became a successful Internet ancestor, while phone books remained in print.

The principle of the Minitel relied on the placement of so-called dumb terminals alongside phone lines. A screen and a keyboard were the only peculiarities of the box, which could not be connected to a printer nor store information. The screen itself displayed what is commonly known as *videotext:* text without images. The French telephone company, which was undertaking substantial improvements to its network, felt that the only way to convince phone subscribers to accept what looked like a tiny computer terminal alongside their phone was to make it available for free. Thus, the fact that the French phone was a monopoly at the time allowed it to reach practically all households who had phones and wished to be connected.

The story would have ended here had there not been an interest in working with private companies eager to capitalize on the new tool and make it into an electronic toy, too. The screens of a Minitel cannot store pictures, only text. Yet, many companies started offering the possibility of early chat lines for groups, information exchange, recipe services, and flight departure information among others. The French sex industry also capitalized on this by offering its own Minitel version of the more notorious phone sex.

The success of Minitel, as several observers put it, was to have brought the information age into the average French household at a time when computers were hardly known and remained expensive. Its success was such that other European nations emulated the principles using television and cable connections. The name varied (in Switzerland, for example, it was known as *Teletext*), but the services were similar, except that no private services were available, only public information.

Ironically, though, the cost in the long term was a lack of interest on the part of both consumers and producers of Minitel to adapt to new conditions, especially at the beginning of the Internet age. The world wide web did not completely displace Minitel, however. It continues to be accessible either through dedicated dumb terminals or online.

CONCLUSION

Of all the means of communication that grew in the modern era, radio was perhaps the most successful in both advanced and developing countries in Europe. By World War II, on average half of both rural and urban populations were connected to voice, and the vast majority by the 1950s. The success was a combination of private initiative, governmental legislating, and most importantly private enterprise that made receivers available at more affordable prices. No other technology, with the exception of the automobile, had such a widespread impact on everyday life. Today, Europeans still pay more attention to radio than any other media. Variations in listening preferences do exist of course. Italians most like listening to car radios (47%), while Bulgarians appear far less interested (3%). Radios at work draw only 7 percent of Spanish listeners, but over 25 percent in Poland and Austria (Cavelier 120). In fact, it is through emulation of personal radio sizes and entertainment features that other tools may end up overtaking it; the cell phone comes to mind. Overall, though, radio (and its cousin, television) introduced a new kind of sociability that later tools have now removed. Instant, but distant, communication is changing the realm of personal relationships.

7

THE SPECTER OF WAR

Up until World War I, despite the evolution of weaponry into greater lethality, the process of conflict was understood as a series of battles, often interrupted in the winter to either allow soldiers to return to their farms or to resupply, as armies could not live off the land then. Mercenary armies were known to have broken for lunch and to collect their dead. Observed tradition called for leaving the civilian populations as they were because new land, in theory, was to be exploited. The reality often did not bear out the practice, though a kind of moral endeavor did exist among officers.

The Napoleonic wars of 1804–1815, themselves a continuation of the revolutionary ones that opposed France to most of Europe, witnessed strategic and tactical innovations, but the practices were mostly the same as before. One notable bit of progress came in the realm of applied science. Napoleon had seeded several scientific endeavors (see chapter 9). This particular one involved a reward for a scientific or technical discovery that would help the French army win wars. The person who earned the reward had spent his life bottling champagne. But Nicholas Appert was versed in the practical usage of chemistry (atomic theory was still being formulated) and had discovered through experiment a means to preserve food. Armies often were slowed down in their momentum by the need to forage for food. After boiling some greens, Appert placed them in a glass container and sealed it. Three weeks later, he opened it before a scientific committee which, after consulting with the French emperor, awarded him the prize. The process allowed the Imperial Army to move much faster than any of its enemies, and when some soldiers were captured with bottled vegetables, the British, thanks to the efforts of Peter Durand, duplicated the process. Durand, a mechanic, found that an even more efficient means of preservation was to enclose the food in tin canisters: canned food thus appeared in British Royal Army bivouacs. It would take until the midcentury, however, before the need to supply urban populations would prompt grocers to store canned items.

There were few radical breakthroughs for technology: It experienced mostly improvements rather than innovations. In the late nineteenth century, guns would go through a radical evolution in the form of the breech-loader. Even particularly bloody scenes, such as Solferino in 1859, were largely overlooked, though the latter's high casualties (40,000) prompted Swiss businessman Henri Dunant, who witnessed the carnage, to create the International Committee of the Red Cross. In the European colonies alone, there were hints that new technologies, such as the machine gun invented by Englishman Hiram Maxim, would have horrible effects on a European battlefield.

Technology fascinated, but its lethal effects were poorly understood. The last major war on European soil had ended in 1871. However, technology was one of the factors in the outbreak of World War I, as each side sought to best the other in devising new weaponry, and glorified it, too.

In particular, the advent of early submarines, but also of *dreadnought*-type ships, suggested an impending change in warfare. The dreadnought was a British design that featured heavily armored ships with heavy-caliber guns. When the first dreadnought was launched, the British assumed they would have the lead, but Germany, which was the primary challenger of British supremacy at sea, soon followed suit. Throughout this early arms race, the public was kept informed of the state of affairs through parliamentary debates, news coverage, and media illustrations suggesting an impending confrontation between the British and the Germans. A navy league in Germany channeled a lobby of businessmen and militarists, but it was also at ship launching gatherings that one saw the importance of the technology as a symbol. In Britain, the ships would be named for nations under the control of England. In Germany, German military hero names usually ended up on hulls. In both cases, the naval ship was a symbol of unity, and all the official speeches were careful to emphasize this (Rüger 140). Other events also convinced populations of the likeliness of future war and of the role technology might play in it.

By 1909, a war psychosis pervaded Great Britain, which included claims that German airships had landed on British territory and dropped off spies. A populist cry for an accelerated dreadnought-building program stated "we want eight, and we won't wait" (Royal Naval Museum). In fact, the Germans remained behind the British in dreadnought numbers partly due to budgetary struggles. Nonetheless, it was clear that the rattling of sabers was more insistent than it had been in decades.

Nowhere perhaps was the immediate impact of technology more visible than in the battlefield over the course of two centuries. Even wide-scale carnage in the manner of Napoleonic engagements could not differ more from a trench assault in 1915. The colorful uniforms, the cavalry charges, and the rather compact shock of men still suggested an element of humanity was associated with wide-scale killing. This would slowly evolve over the course of the nineteenth century and prompt sudden changes in World War I.

Technology is not the only factor to account for these changes (economics, politics, and culture all play a role). Take, for example, a major innovation in military technology, the breechloader. Using magazine clips to load and firing with the use of smokeless powder, the breechloader became the standard weapon for all major infantry units in the 1880s. Prior to that, several armies had made use of it, and many observers were tempted to argue this explained early victory in some conflicts. However, tactics and strategy, often based on proven practices, seemed to make the difference in victory. That said, Europeans were afforded a hint of the horrors of the twentieth century when they read reports of such battles as Solferino in 1859 (which prompted Swiss Henri Dunant to found the International Committee of the Red Cross) or even learned of free fighters (the precursors of guerilla warriors) in the Franco Prussian war of 1870.

A major new invention, the Sir Hiram Maxim machine gun, did see usage and give a tactical advantage to its owners in overseas battles. In Africa, for example, relatively small European troops were able to destroy indigenous armies by mowing down spear-armed attackers in a matter of hours. But the butchery that occurred overseas was simply recorded as a victory rather than a warning of what similar assaults could do on the European theater.

Commentators often point out that generals fight the previous war. Though simplistic, the argument contains a kernel of truth where World War I was concerned. Tactics change slowly not only because of strategy and goals but also because it takes time to retrain troops to deal with new threats and new technologies. In the case of European armies, much of the knowledge of war was built in the eighteenth century, at a time when neither economics nor technology were considered important factors in war. This remained the case throughout the nineteenth century, even as new weapons began to appear When technology did matter, it concerned specific logistical issues. For example, many European militaries became convinced of the necessity to school all men so they could read and write, partly because the technologies they would be called upon to use were complex enough to require some form of schooling before field practice. Similarly, the railroad was considered essential to assist in troop movement. The Schlieffen attack plan, which was Germany's blueprint for an invasion of France in the years leading up to World War I, called for the gathering and movement to the opening front of 3 million men using

The Balkan War of 1911–1912 was but a prelude to World War I. Here, Italian troops maneuver their cannon. The use of long-range artillery as well as the first dropping of grenades from airplanes did not draw much attention, even though it heralded what would happen on a huge scale. Source: Propaganda postcard, author's collection.

11,000 trains over a two-week period, thus beating France by a few days. The theory, however, quickly slammed into the practice of technological warfare, whereby millions of men would experience a conflict like none of their forefathers had faced before.

World War I, also referred to by its contemporaries as the Great War, quickly transformed itself into a conflict unseen before, not just in scale, but in its nature. The opening moves suggested if not enthusiasm, at least resolve among the belligerents, with a certainty (emphasized by their respective leadership) that the capital city of the enemy would be occupied by Christmas. This presumed that the war would be fought along the lines of Napoleonic grand-scale warfare involving major attacks; it quickly ground to a halt in the face of massive resistance on both sides to offensives and counter offensives. The machine gun, by then half a century old, had become the defensive weapon of choice, annihilating entire battalions in seconds. By October 1914, on what became known as the western front, was also the first total war.

The main European nations' full resources of military, civilian, economic, and political power were thrown into the conflict. Consequently, the fabric of society changed. When German industrialist Walther Rathenau sought to warn a general in 1914 that supplies would run low fast, he was met with skepticism, but he soon convinced General Falkenhayn, the Chief of

Staff, that without a complete reorganization at the rear there could be no drawn out war. The home front, as it became known, introduced military technology away from the lines by placing civilians, notably women, in shell factories but also by exposing them to bombings by long-range cannons and early bombers. Naturally, though, technology's ugly war impact was felt at the front first and foremost.

The car and truck were immediately put to use in the Great War. Logistical support came to include not only kitchen trucks but also mobile chapels. In fact, as the war to end all wars entered its first winter, reports often included how valuable the car had been. Though seemingly a given, it was in fact unclear a few years earlier whether this would actually be a useful tool.

France, England, Italy, and Switzerland all became rapidly aware of the future organization role the car would play in the army, but no one could find a solution to the important problem of fuel supply. In the case of a conflict, the reserves, disseminated in different countries, would soon be depleted and the nonproducing countries could find themselves in dire straits. In 1905, in Germany, Prince Heinrich of Prussia, brother of the German Emperor Wilhelm II and president of the German Automobile-Club, created a unit of military drivers who, in case of conflict, had to make available their vehicles and their mechanics to the army.

From 1905, Austria produced the first modern armored vehicle. This four-wheel drive weighed two tons and was equipped with two machine guns. Yet, the imperial army wondered whether this new weapon should be developed because it frightened horses, which trembled uncontrollably when it was driven past them, back-firing as it went. The idea was dropped, only to be reinstated 10 years later at the height of the Great War.

As for the Swiss army, it created a volunteer drivers corps in 1907 similar to Germany's. It comprised 130 volunteers "of Swiss nationality, deemed fit, both physically and technically, devoted to the automobile and able to own and keep a car between 15 and 35 horse power constantly maintained" (de Syon and Sion 47). In times of peace, the cars were mostly used to transport staff officers. However, the efficiency of this corps of drivers, created in peacetime, would prove very limited in times of war. It became necessary to employ soldiers as drivers and professionally train this army corps. The surplus materials, both cars and drivers, were sent back to civilian life.

On June 28, 1914, Austro-Hungarian crown prince Archduke Francis Ferdinand was in a car at Sarajevo on his way to observe army maneuvers. Having taken the wrong route, his driver ended up at the very spot where an assassin was awaiting the royal couple. Two bullets. Two deaths. The black car entered the history books, carrying the two dead bodies to the hospital crumpled on the back seat. They were the first two victims of a war that was to claim almost nine million lives. The car was preserved in the Viennese military museum, where it still sits today.

In September 1914, the German offensive was devastating, and Paris appeared close to falling. The commander of Parisian defenses, General Galliéni requisitioned all 600 Paris taxicabs to drive five infantry battalions, rested and ready for combat, to the front, 31 miles away. Four thousand men do not represent a massive contingent in the midst of a million soldiers, but the psychological impact of the Taxis of the Marne on the morale of the enemy troops was substantial and helped stop the Germans.

To solve various logistic problems, limousines and touring cars were stripped of their sumptuous bodyworks and their luxurious leather interiors. Elongated and barren, the chassis now carried a wooden box and side rails. Often, a trailer was attached to the older cars, reducing them to a utilitarian role. Little by little, the automobile would become a machine that kills, dresses wounds, and feeds mouths. Even on the front itself, these motorized vehicles provided back up and solved all kinds of problems: workshop trucks, postal trucks, communication centers, mobile kitchens, wire-cutter cars, trench diggers' ploughs, distillery trucks supplying boiling water to prevent epidemics, car chapels, not to mention the use of ordinary cars, simply equipped with metallic shields instead of windshields and lateral steel plates, used by patrols.

At the end of 1916, allied newspapers spoke with enthusiasm of the services rendered by the armored cars during the last offensive on the Somme: "This admirable progress in the art of destroying the enemy would have nothing admirable about it had there not been a counterpart of life on the side of the one who uses it. It also enables one to glimpse at a new field of activity for the car in the peaceful conquest of the world. So many new countries, inaccessible due to lack of roads, will be rapidly transformed and become sources of wealth. After the work of death—the work of life" (de Syon and Sion 52).

The war affected automobile production, as factories and workshops underwent conversion. Everywhere, women replaced the men fit for war or already at the front. These men viewed the car with a mix of appreciation (when they could board it) and circumspection. Indeed, it is in the trenches that men first realized that warfare had changed because of the technology involved.

The most immediate experience of conscripts and volunteers with technology and warfare involved the gun, but in the field, two other unlikely tools became part of everyday life: the shovel and the barbed wire.

The shovel became a standard part of infantrymen's equipment as the war moved quickly in fall 1914 from a shifting front to one of entrenchment: digging was part of the daily routine, either to reinforce positions, fix what had been bombed the previous day, or to try and burry bodies. The latter function became nigh impossible as the no-man's land separating enemy trenches saw the piling of cadavers ever higher.

Though the machine was the primary cause of mounting casualties, the barbed wire also changed tactical warfare. An invention of the nineteenth

Requisitioned buses were quickly put to use in fall 1914 to carry the wounded away from the front. Eventually, specialized ambulances, including some driven by American women volunteers, would assume the job of moving the wounded from midstations to train stations or directly to a city. Source: Propaganda postcard, author's collection.

century, barbed wire is associated in American minds with the fencing of the prairie to herd cattle more efficiently. By September 1914, barbed wire was in place in German trenches, and soon after the French and the British followed suit. It had remarkable defensive advantages. Of light weight, it could be deployed in various forms (accordion style, or sometimes as an extended mattress-shaped rug) yet remain in place during artillery bombings. As soldiers attacked the other trenches (going over the top), they sought to take advantage of the lull that separated the artillery barrage from the enemy's machine gun response. Yet, as they reached the other trench, they would become tangled in barbed wire, offering themselves as targets to the guns of enemy soldiers.

Time and again, various attempts at ridding the field of deadly barbed wire failed. Artillery barrages could not blow them away, though with luck, a targeted explosion might bury a section in mud. Several gifted officers and soldiers devised informal methods to either cut through the barbed wire (it was often done at night by designated "volunteers" who crawled through no-man's land with big pliers), use a British-designed Traversor (essentially a very thick blanket that would cover the barbed wire so one could crawl), or push wooden planks over sections to walk over (two or three men had to do this, thus exposing a bigger target to enemy fire). Eventually, the tank, first introduced by the British, would

French troops lay phone wires amid the barbed wire above a communication trench to prevent attackers from jumping into it. Source: From a soldier's memento album, author's collection.

offer some respite, though its success varied according to the terrain and strength of enemy artillery. Regardless of methods used, the barbed wire remained a deadly symbol of trench warfare. Its horror consisted of rotting corpses, which so demoralized some men that they risked their lives to try and unhook them (Razac 49).

Things never got better. The war of attrition each side fought was characterized by tactics involving traditional warfare (calling for an all-out attack) but encountering resistance. On the first day of the battle of the Somme on July 1, 1916, British troops were told to prepare for a "walk" across the lines following several days of artillery shelling of the German positions. The shelling, however, had also sent the Germans a message that a full attack was pending, and they folded back to other positions during the shelling. Though shaken, German defenses redeployed when the shelling ceased, and the machine guns sputtered as soon as the British went over the top (of the trenches). Within five hours, the British had lost almost 60,000 soldiers, killed or wounded. The battle lasted another five months and ended up causing 419,000 British casualties and approximately 200,000 French and 500,000 German losses. All for an Allied advance of six miles.

New weaponry failed to change the equation. To try and break the deadlock, each side introduced or perfected several new weapons, none of which offered a decisive advantage. These included airplanes and airships, gas warfare, radio communication, and the tank, all of which complemented long-range guns.

The flying machine was young when war broke out, and though its psychological impact was substantial on several levels, it remained a peripheral weapon whose full potential had yet to be realized by the time the conflict ended. High commands on all sides had only recently realized that, contrary to general Foch's 1911 observation that "airplanes are interesting toys with no military value," the machine could in fact help. Hastily constituted air forces, however, initially had only one purpose, observation. The airplane replaced the blimp on some theaters, as it was a more difficult target to aim for. Pilots sent aloft by general Joffre in September 1914, for example, provided information on German troop movement that allowed the shifting of troops toward the Marne, thus stopping the German march on Paris.

Pilots were very much at the mercy of their machines, however. Testing was still in its infancy, so actual awareness of an aircraft's strengths and weakness often occurred in the field. Some pilots became victorious accidentally, when pulling out of a steep dive where their enemy failed to do so, or succeeding in a particularly daring loop that the other aircraft did not have the capacity to carry out. Engines were unreliable and sputtered oil and fuel into the operator's face (which is why many wore long scarves to help wipe the goggles clean). The cold forced them to wear very heavy gear that prevented easy movement within the cockpit. Furthermore, with the exception of the Germans, none wore parachutes. The Allies considered these a cowardly device, unworthy of consideration. In so doing, they contributed (as did the pilots, who generally accepted such arguments) to the myth of the flying knight that would develop around fliers and later aces.

The idea of shooting down enemy aircraft came about very quickly, but there were no means to do so. Two-crew aircraft would have a side arm or even a rifle on board so that the observer might try to shoot at the enemy. This rather outdated approach reminded some of a duel, and the behavior of pilots tended to confirm this. Once machine guns were installed on board planes (by 1915 it was a standard equipment), the kill ratio changed substantially. Yet, chronicles of that era confirm that, at least in the first two years of the conflict, pilots on all sides broke off combat if they noticed their adversary's guns had jammed. Similarly, pilots downed behind enemy lines were given special treatment quite different from that of foot soldiers. Finally, pilots often would fly to the area where they had registered a kill and dropped a memento, usually a wreath honoring their dead enemy.

Adding to this chivalric construction, the French air force introduced the identification of "ace" for any pilot who had downed at least five enemy aircraft. The practice quickly spread to other air forces and continues its existence today. These aces became the stars of the war effort, paraded at home and pictured in newspapers. Their actions, though paling in comparison to what "trench grunts" underwent for months at a time, cast them in a role that matched more closely popular visions of a heroic war, the last of which had been fought in the preceding century. Technology on the ground dehumanized men and made killing impersonal. In the air, however, it succeeded in maintaining the illusion of honor and braveness as each side tried to down the other. Though much of the actual chivalry disappeared by 1918 (partly because most great aces did not survive the war), the aura remained, transformed into myth.

Another reason the aura eventually disappeared was because of the flying machine's slow transformation into a strategic bomber. In the opening weeks of the war, aircraft were first used as observers, though some German flights over Paris did involve dropping a grenade and some propaganda leaflets. Very quickly, however, the notion of hitting a target in the rear, became attractive. In the context of a total war, where industry was gearing towards a long-term conflict, hitting factories in the rear seemed a legitimate strategy. Zeppelin airships thus became the first strategic bombers. Their success rate was extremely limited, but the fact that they could fly long distances made them formidable weapons. Londoners and Parisians, not familiar with such machines, often tried to see them when the alarm sounded rather than hide in the basement! It took several bombings, and the resulting horrors (civilian houses were often hit due to poor navigation techniques) slowly taught Europeans to fear the skies. Indeed, by 1917, a series of bombers had been developed on both sides. Western German towns experienced the terror of aerial bombing, too.

The advent of gas warfare represents an important case of science applied in warfare. Fritz Haber, the inventor of the ammonium nitrate synthesis (see chapter 1), led the Kaiser Wilhelm Institute for Physical Chemistry in Berlin. When the war begun, he contacted the German government and placed his laboratory at its service. His research led directly to the development of poison gas as a weapon. After being tested near Cologne in 1915, the gas was declared operational and deployed against French troops in the spring.

The impact of such an attack was stunning. As the *New York Times* reported on April 26, 1915:

> The French soldiers were naturally taken by surprise. Some got away in time, but many, alas! not understanding the new danger, were not so fortunate, and were overcome by the fumes and died poisoned. Among those who escaped nearly all cough and spit blood, the chlorine-attacking the mucous membrane. The dead were turned black at once (p. 1)

The Allies quickly devised chemical warfare of their own but found it was a very difficult weapon to use. Whenever the winds turned, soldiers who had unleashed the attack risked death themselves. The Germans, too, were dissatisfied with the weapon. As soldiers breathed the chlorine gas, they coughed, which actually limited their intake of poison. Furthermore, the gas would clear after 15 minutes, which meant fresh troops might come in and replace the dying. What was needed, from the vantage point of the tacticians, was a gas that would not disperse so fast and would actually hurt the victim without even being breathed. Haber went back to work and eventually devised a phosgene gas that remained in suspension as droplets (in the manner of a fog) and attacked the skin of the victim, resulting in bubbling skin. Breathing it actually drowned the soldier as his lungs filled with water to try and remedy the blister the gas caused. Popularly known as "mustard gas" because of a smell that resembled the condiment, phosgene were eventually placed in shells, too. Yet, as each side devised procedures and methods to deal with the threat, gas warfare, like other new weapons, proved unable to break the stalemate.

The impact of total war affected not just *what* was manufactured but also *who* made it and *how*. In aviation in particular, skilled craftsmen who had worked on airplanes in a quasi-artistic fashion saw the importation of scientific management methods from the United States. This is all the more notable for aside from mechanics and engineers, the main skilled workers were cabinet and furniture makers in an industry that converted its production to aircraft cells (Edwards 74–75). The shift towards supply controls, single activity in a chain assembly, and standardization of blue prints occurred under war emergency conditions and would have important implications for the postwar world.

The destruction of many urban areas through strategic bombings paradoxically accelerated the pace of urban renewal, yet, in so doing, it also changed the nature of the city and of the countryside. The urgent need for housing meant that buildings went up with limited attention to modernization. The world war, though not necessarily technology, also accelerated the growth of cities because of the decline in agricultural activity.

MEDICINE AT WAR

The horrors of World War I changed the way medicine was practiced on the battlefield. Not only were men wounded by the thousands, thus straining the logistics of evacuation and care, but the nature of their injuries was unheard of. Gas, flame throwers, and random shelling all maimed, yet, the degree of early survival was equally hard to fathom. Many a surgeon would struggle with his patient because the kind of wound the man suffered had never been mentioned in the medical books. Disfigurement in particular proved the greatest challenge. Men survived torn-off jaws and burned faces but were de facto barred from civilian life for none of their loved ones would recognize their physical features. A member of the British Medical Corps described such wounded as "gargoyles."

The effect of disease should also be considered. There was high demand, for example, for compounds that would avert typhus and tetanus (penicillin did not exist yet). While this encouraged the development of a pharmaceutical industry, it did not avert the spread of illnesses. In fact, in the last year of the war, two-thirds of British casualties were from disease as opposed to combat wounds. Many soldiers, weakened by months spent in insalubrious trenches, ended up victims of the Spanish flu (see chapter 10).

Generally, World War I forced research upon the medical establishment, as it strove to treat so many new kinds of wounds and associated diseases. Yet, the war failed to encourage substantial investigation of mental health. The culture of Europe remained patriarchal in nature, and social mores would not tolerate crying men. Yet, shell shock, as it was known early on, had clear physical manifestations including hysteria, confusion, paralysis, and loss of speech. It would take several decades before it became accepted as posttraumatic stress disorder. The reason for dismissing the symptoms was simple: Soldiers broke down, refused direct orders, and sometimes ran away. The treatment, when it was available, often included electroshocks. There were some psychiatric treatments that involved talking the patient through his trauma, but what is nowadays called cognitive therapy also did not have many followers.

LEARNING FROM WORLD WAR I

The war to end all wars fooled nobody. In fact, military and civilian populations alike kept fresh the memories of the conflict. High commands now understood that the concept of total war was one here to stay. Military thought generally evolved along the lines of quick warfare using aviation and tanks, or position warfare.

In the case of position warfare, the years in the trenches affected French military thought considerably. Throughout the 1920s, the French General staff argued about what the best defensive attitude was and where to build fortifications. Eventually, the focus fell on the relatively flat area that extended from the Ardennes to Alsace. In this context, politician André Maginot came to oversee the decision to create a defensive line that bore his name. A war minister from 1922 to 1924 (and again in 1929), he remained involved in the planning as the head of the Parliamentary Armament Committee. By the time the Maginot line was built, it had cost twice the original estimate, exceeding six billion French francs at the time.

The design of the Maginot line posed several challenges to engineers, and no two forts (also known as *ouvrages*) were the same due to terrain and communication constraints. In addition, drainage was a constant worry, to the point where several modifications were required in the 1930s to make the installations livable. Common to all forts were the garrison personnel, ranging from 200 to 1,200 men, and divided into infantry, artillery, and engineering sections. The layout followed the same structure everywhere, with living quarters at the bottom of the installation, a railroad that brought supplies to a narrow gauge network that linked forts, and a diesel-powered factory to supply electricity in times of war (in peace time, the French national power grid fed the entire Maginot line).

The Maginot Line represented symbolically a sense of fatalism regarding warfare. On the one hand, the line might hold the enemy in the manner that fortress cities had since the Middle Ages. But the presumption was that attacks would be so brutal that one had to hide underground. Indeed, perhaps the most important side of war's impact was the general notion that science and technology could no longer just do good.

WORLD WAR REDUX

To European contemporaries, the only thing that distinguished truly World War II from its predecessor was the scale of violence. In fact, technologies seemed to have simply evolved to become more lethal, and while there were of course new tactics and strategies adopted and tested, the impact on everyday life was felt as more of the same, only worse. Advanced technology, in the form of better airplane range, stronger engines, improved aerodynamics, and navigation all made strategic bombing a standard feature on all sides, as cities were leveled in a matter of days, or even a single night. As many as 600,000 civilian Germans died, and 90 percent of all urban housing in their country disappeared. The total war approach had been applied, and the ultimate strategic bombing, a nuclear one, destroyed Hiroshima and Nagasaki in August 1945. This latter fact would change the equation of warfare for ever. However, World War II also introduced another, new horrific concept: industrial-scale genocide.

The notable exception was that gases were not used against armies but against civilians. It would be difficult to summarize the horrors of the Holocaust here, but two facets of it, namely industrial-scale destruction and medical maltreatment, should be cited. Indeed, taken together, both contributed to the ultimate removal of the notion of science as a neutral and objective field. The two facets intertwine, and it is with the racial construct the Nazis established as early as 1933 that we begin.

Nazism was based on a biocentric vision of Germany, whereby any one deemed undesirable on either racial or political grounds was to be excluded from the People's Community, a loose matrix referred to in propaganda as the Aryan race. Early steps toward exclusion were based on the eugenic practices already advocated before Nazism in some American and European circles. Not surprisingly, then the Nazi regime took steps in its first year toward exclusion at the biological level. While Jews were the obvious target, the very fact that they were no longer considered Germans meant other groups still German would be targeted first. A sterilization law was thus passed in July 1933 authorizing the sterilization of criminals at judges' discretion. The primary targets in this case, however, were homosexuals. Other measures included anthropometric measurements to determine one's exact racial make in all cases where Jews were not involved. In the latter case, however, the Nuremberg Laws of 1935 completed legally the exclusion of Jews from German society by identifying who was Jewish on the basis of blood relations, be it "full" or "mixed."

Yet, this process of legal exclusion, while it did include violence, was not systematic extermination. It is in the context of World War II that the tragedy would come. On September 1, 1939, as Germany attacked Poland, Adolf Hitler sent a note to a physician from his entourage, tasking him with the organization and implementation of euthanasia in Nazi Germany. Euthanasia involves the right to die but implies that the patient requests the help of a physician in dying. Here, the so-called mercy death was hijacked, as the physician alone would be allowed to decide when a patient would be killed. The operation, a secret plan coded T4 (for the initial and number of the Berlin street where it was managed), would last almost two years. Its primary targets were mental patients with no family, elderly in limited contact with their families, and eventually children with low IQs. Perfunctory examinations allowed the staff to identify the patients to be terminated, and these would be taken to specially designated execution centers in Germany (not concentration camps). There, after another exam, death by CO_2 inhalation would follow, and the body, after autopsy, would be cremated. This last measure was meant to ensure that any surviving family member could not ask for a postmortem exam that would show the victim had not suffered a heart attack, stroke, or even a falling accident.

The exact number of victims is hard to ascertain but conservatively reaches at least 90,000, many of whom were children, some with simple learning disabilities. Hoping that the war emergency would dissuade family members from asking questions, the Nazi authorities were surprised when they encountered symbolic forms of resistance. Death notices, usually following protocol, now included comments like "incomprehensible," or "no condolences accepted" (Friedlander 177). Reports from the secret police showed the general population as restless, and several priests and ministers began protesting such actions during Sunday service.

Consequently, in late summer 1941, T4 was shut down. It migrated in another form into concentration camps, but as several historians have noted, some of the T4 experts would then become centrally important when the Nazi genocide actually started.

The actual Holocaust began in summer 1941, when Nazi Germany invaded the Soviet Union. By then, thousands of Jews had already died in the horrible conditions of ghettoes set up in German-occupied Poland. In the case of outright execution, though, several technical solutions were tried. Systematic shootings were the first practice, of course, but the scale of executions made it difficult to complete (all bodies had to be buried, and the executioners themselves were not always able to withstand the pressure of shooting civilians). An attempt at a more efficient method involved the use of gas vans, where a hose connected to the exhaust pipe was linked into the sealed rear area of a truck. In both cases, the tools remained in use for years but were not deemed efficient. The perpetrators' motivations suggest a variety of grounds for their actions, of which anti-Semitism and following orders were not necessarily the top ones. Memorandums do suggest an almost obsessive measure of professionalism. In the following excerpt of a report on the malfunction of a gas van (the victims were not dead after the van had driven around for 30 minutes):

> A reduction in the load area appears desirable. It can be achieved by reducing the size of the van by c. 1 meter [3 feet]. The difficulty referred to cannot be overcome by reducing the size of the load. For a reduction in the numbers will necessitate a longer period of operation because the free spaces will have to be filled with CO. By contrast, a smaller load area which is completely full requires a much shorter period of operation since there are no free spaces. (Just)

The concern for the victims is nonexistent. All that counts is to get the job done, in this case fix a malfunctioning truck. A similar approach stands behind the design, construction, and operation of gas chambers.

In the concentration camp system, six installations functioned specifically as extermination camps. The most notorious was also the biggest: Auschwitz. This is where Zyklon B, an insecticide, was first tested in September 1941 on Soviet prisoners of war. Whereas it could take three days to destroy lice in a sealed up house, Zyklon B, an industrially manufactured granule compound of hydrogen cyanide, could kill a human in a few minutes. Consequently, thousands of victims, primarily Jewish, were killed in the gas chambers of Auschwitz. To destroy the bodies, an industrial design proposed by the Topf Corporation involved the operation of multiple crematoria. Jewish prisoners moved the bodies and operated these. By the time the Soviet army liberated Auschwitz in January 1945, as many as 1.1 million people may have died by this industrial process. Some 5.9 million Jewish and Romani civilians were eradicated using some of the methods outlined previously. Some were tortured, in particular in medical experiments (see chapter 10). The horrors associated with this misuse of technological and industrial knowledge continue to fascinate and are also the subject of interest of fields beyond history.

THE IMPACT OF THE BOMB

The other symbol of mass death associated with World War II involves the dropping of the atomic bomb. Though this was an American action, the gestation of the weapon was the result of a series of research endeavors, most notably in England, France, Germany, and Italy over the decades that preceded World War II. The Manhattan Project, as it came to be called, benefited from collaboration in the early stages between British and American scientists. And while the actual bombing of Hiroshima and Nagasaki stunned Europeans, for whom the war had ended three months earlier, there was little discussion of what the nuclear age actually meant. As relations between the United States and the Soviet Union deteriorated, however, and especially once the latter acquired the atomic bomb, too (in 1949), more attention focused on the implications of the bomb. Two U.S. allies in Europe also acquired nuclear weapons: Great Britain in 1952 and France in 1960. The result was that Europe found itself at the center of a strategic nuclear game.

In the context of the cold war, the risk of nuclear war became a central feature of everyday life due to the very fear it instilled. The paradox is obvious: considerable time and money were expended to deal with a technology that fortunately was never used.

In Switzerland, all new buildings were required to have their basements constructed as nuclear shelters. Heavy cement doors on reinforced steel hinges were in place, combined with emergency exits, and air filter apparatus were installed. Within the protected areas, the standard cellar divisions made of wood were in place but constructed in such a way that planks could easily be nailed to the sides and turned into bunks. In case of attack or simply of an alert, Civil Protection units would travel to all apartment buildings and make sure everything ran properly, and they would appoint a block leader.

In theory, the system was remarkable and elicited the envy of most specialists in civil defense. Actual tests were conducted in the 1980s and filmed for a TV documentary. The task was to remain below ground for 24 hours. To avoid too much difficulty, only neighbors who knew each other in the selected apartment building were picked, and all preparations were completed without time pressure, contrary to what might happen in case of nuclear crisis. The result was far from successful. Even with plentiful supplies, working equipment, and radio contact with the outside world, the social and psychological impact of being cooped up together was substantial. After the first 12 hours, which involved excitement, jokes, and board games, friends in the open turned into aggressive co-survivors, arguing about matters as mundane as sound, the weather upstairs, and, of course, politics. The prognosis based on social and psychological analyses was that survival in such conditions for two to three weeks (the standard expectation to ensure maximum dispersal of the fallout radiation effect) was nigh impossible. Still, in light of the lack of alternative, civil defense procedures continued unabated.

In countries where civil defense shelters were insufficient to house the entire population, there was either no discussion of what to do in the event of an attack or genuine attempts at informing the population of such issues. England is perhaps the best example of the second approach.

In the 1960s, the movie *The War Game* was prepared for broadcast in a Wednesday weekly series on the BBC. Scheduled for release on August 6, 1966, the 22nd anniversary of the Hiroshima bombing, it was pulled from the program and was not released until 1985. Its producer, Peter Watkins, left the country in protest. When asked, the BBC replied that the discussion of horrible survival conditions was too difficult to broadcast for a lay public. Though fictional, it was based on governmental studies of what would happen if a Soviet nuclear attack hit Great Britain. It received an Oscar for best documentary.

By the 1970s, though, the threat of nuclear attack seemed further confirmed as relations between the Soviet Union and Western Europe deteriorated. The signature of several treaties on nuclear weaponry did little to calm fears. Consequently, Great Britain also implemented a "Protect and Survive" program seeking to inform the population of what it could do to protect itself in case of an attack. Though the advice was based on expert opinion, it nonetheless gave the impression that an actual nuclear bombing, though formidable, would be akin to a World War II bombing campaign. Investigative journalists later showed that such things as the building of temporary shelters in home would only work if one were at least 7 miles away from the epicenter of a one-megaton explosion. The BBC undertook to address the issue by commissioning a series of 20 short documentaries (about three to five minutes each) on what to do in case of attack.

When governments did not discuss the bomb, popular culture stepped up and faced the nuclear nightmare. English writer Nevil Shute, an engineer by training, wrote one of the classics of the genre, *On the Beach*, later made into a movie (1959).

The War Game concept was revisited in *Threads* (1984), which considered the life of a family after a nuclear war yet followed the documentary style of *The War Game*. The movie made regular reference to the documentaries of the "Protect and Survive" series up to the actual attack, and in so doing, suggested the futility of the protective measures one might take. Even surviving meant little if society began to fall apart afterward.

Comics, too, took on the matter of nuclear war. Some did so in the context of a dystopian scenario, but some tried to discuss the actual impact on civilian life. Raymond Briggs, an acclaimed British author of children's books, took on the matter in *When the Wind Blows*, which followed a retired couple in the countryside and how they tried to survive based on previous experience, namely the World War II blitz. The book was so successful that it was optioned and made into a cartoon in 1986.

In the meantime, substantial popular demonstrations against the bomb occurred around Europe, most notably in England and Germany, where new American nuclear weapons were deployed in 1983. At the Royal Air Force base of Greenham Common, starting in 1981, a group of pacifist women set up a camp intended to protest the presence of nuclear weapons there. On April 14, over 30,000 protesters joined them in a human chain around the base. The weapons were eventually removed following the Intermediate Nuclear Force reduction treaty signature in 1987. Though the women were often described as manipulated idealists, they were successful in focusing the nuclear freeze movement and, in so doing, inspired emulations of their movement, notably in West Germany. Furthermore, regardless of the effectiveness of their protests, they reflected clearly the European unease with military technology after having experienced two world wars on its soil.

CONCLUSION

A popular poster on sale in western Germany for years displayed a poem in local dialect, which translates roughly as follows:

When my grandmother was a little girl, there was war.
When my mother was a little girl, there was war.
When I was a little girl, there was war.
When my daughter was a little girl, there was peace.
Hopefully this will hold for a while!

Allegorically, this limerick summarizes the European experience with warfare and its associated technologies. Europe experienced two world wars on its soil and preparations for a third one, too. As it did so, its culture shifted. The antiwar protests grew, and in the post–cold war era, it is hard to dismiss protesters as left-wing idealists, as was commonly done for years. The European ambivalence about the use of force nowadays grounds its origins very precisely in the trauma industrial warfare caused.

8

ENERGY

Over the course of two centuries, Europeans were drawn to the power of steam, coal, gas, fuel, and nuclear power. Several of these were central to the manufacture and use of electricity, but as we shall see here, the dependence on a particular energy could not be assumed. The assumption that a generalized technology came in response to demand overlooks the fact that consumers have to be convinced that something new is useful to them. Gas and electricity producers experienced this as they promoted their services. The automobile was not welcome at first, and nuclear power to this very day elicits an ambivalent response. Several of these energies are discussed elsewhere in other chapters, which is why the primary focus will be on coal and on nuclear energy.

COAL

Accessing coal became a central preoccupation of the Industrial Revolution. Immediately, however, the dangers associated with mines became apparent, especially with firedamp explosions and associated gallery collapses. The most notorious ones happened in Western Europe in the nineteenth century, and Emile Zola chronicled the risks in his novel *Germinal*. Simply put, firedamp consumes itself into huge explosions when a single spark comes into contact with a cloud. Because miners initially used open flame candles to see (the limited lighting also caused blindness), the risk of tragedy was huge. In some cases, volunteers acted as sacrificial lambs by attempting to light the gas using long lit sticks. But generally, engineers moved to implement special safety features by the late nineteenth century that included air vents. Accidents grew rare but were never completely out of the collective memory of miners.

GAS

During the late Enlightenment, several amateur scientists toyed with a discovery from Englishman James Clayton. In 1664, Clayton had noticed that distillation of coal could yield a flammable lighting gas. However, the discovery did not elicit any attempt at replacing candlelight and oil lamps. Over a century later, engineers Frenchman Philippe Lebon and Scottsman William Murdoch discovered independently from one another means of synthesizing lighting gas efficiently. The former used wood, the latter coal.

Lebon patented his discovery in 1799 (though some sources claim it was only granted two years later) but encountered opposition to his lighting gas on three fronts. First was his own profession as an engineer, which took him far from Paris and limited his ability to seek out investors. Second, the people he approached tended to be private citizens rather than industrialists, which meant that funding was slow in coming and limited. Finally, the existing city-lighting industry was based on oil lamps manually lit at night. Fearing for their livelihood, Paris's lightmen mounted protests and emphasized the weakness of the system: the smell the burning gas gave out. Lebon never saw the eventual adoption of his invention. He was killed by street thugs at night in 1804.

Paris eventually followed London by adopting Murdoch's synthesis. The latter had successfully found a way to purify gas, thus limiting the noxious odors that affected Lebon's production. But it was a competitor, Englishman F. A. Windsor, who spread the invention into London and across the Channel. The event is noteworthy because it reflects the shift from inventor to businessman. Windsor was granted a Royal Patent in 1810 to light up parts of London but encountered popular resistance because the gas synthesizers were deemed ugly. Others feared explosions, though these occurred rarely. The end of the Napoleonic wars in 1814, however, became an occasion for night celebration at which gas lighting was abundantly demonstrated. Londoners slowly began accepting the value of gas lighting.

As for Windsor, he sought to market his system in France starting in 1815. Initial opposition soon gave way to fascination as Windsor undertook various demonstrations of his system. Once he received approval of a French scientific commission, however, the French also accepted the arrival of gas lighting. Members of the elite did complain that the color of the flame affected perception, and some even suggested it made women ugly by casting odd shadows on their faces. Gas now accompanied Europe well into the late nineteenth century. It was eventually replaced with petrol oil (notably in Bucarest, Romania, in the 1850s). Gas manufacturers would soon search for other applications for their product (see chapter 9).

ELECTRICITY

Multiple scientific discoveries in the late Enlightenment provided the basis for potential applications of electrical power. Most importantly in the era of interest, Michael Faraday demonstrated in 1831 that mechanical power might be converted into electricity. Yet, it wasn't until 1857 that generators became standard features of such buildings as lighthouses.

Arc lights (the passing of current between two carbon rods) became standard features in cities installing new street lighting. Because the light generated was intense, it was attractive in the public sphere. The Gaiety Theater in London, for example, made use of it starting in 1878. No homes, however, adopted it. This would have involved too great a shift from the mild, even weak, light gas lamps generated to the strong, raw whiteness of an arc light (Hannah 3). Electrical lighting of public areas gave the illusion of festive opulence, something gas lighting could not equate. To have electrical light was to be modern. The fact that electricity was also considered a curiosity added to the effect. In major French cities, for example, public electrification slowly started to spread in the 1840s, but the universal exposition of 1851, while not introducing new technologies, played a major role in spreading the awareness of electrical technology.

Yet, like so many other discoveries, the application of electricity had to be developed and encouraged: Invention was the mother of necessity, and not the other way around. In the case of electricity, the supply side of the industry had to generate new ideas beyond the obvious use of energy for lighting.

In the 1890s, the electrical industry already noticed that its load factor (the ratio between actual electricity supplied and what could have been supplied at full capacity) was relatively low and did not grow at any expected pace. The reasons for this included lack of understanding of the new technology, but also its cost (wiring a while house was prohibitively expensive) and the limited use of electricity beyond lighting. Historian Thomas Hughes summarized the issues well when he spoke of "networks of power": Engineers involved in solving the problem of electrification and also in devising its use developed what one might call national styles. In the case of Great Britain, supply engineers sought to identify what might increase the load factor most and identified cooking and heating tools. The challenge for them was not only devising such items but also making them more attractive than the direct competition, namely, gas. Also, using, say, a toaster and a coffee maker increased the load factor when very little electricity was actually consumed, in the morning (Forty 187).

Early advertisements for electricity made use of fairy-tale characters that played on the seemingly magical aspect of the light bulb. Edison's invention, which quickly spread to Europe, was probably the best exemplification of the energy's new domestication. Source: Astra, Geneva, Switzerland; reproduced by permission.

The adoption of electrical standards was also a challenge. Initial discussion over the use of direct current (DC) versus alternative current (AC) saw the latter win eventually but with a different voltage and frequency. Voltage is relatively adaptable, but the frequency, measured in hertz, indicates the number of times the polarity is reversed per second. AC eventually won out because it could be sent over long distances. However, frequency ended up determining the differences between the American and European continents. In America, Westinghouse was using 60 Hz plants for its electrical lighting, and perfecting converters was done using this frequency. In Europe on the other hand, the main manufacturer of electricity on the continent, the German AEG, used 50 Hz as standard frequency and, thanks to its quasi monopoly, spread the norm across the borders. In the United Kingdom, several standards were kept, partly because electricity was initially the business of private companies. The eventual nationalization of the power grid as well as economies of scale (the need to sell to a wider base of customers) ensured that the 50 Hz standard was adopted by World War II.

The United Kingdom also provides an interesting case study as far as the pattern of adoption of electricity goes. By 1900, with the exception of trolleys, electricity was slow to gain adoption. The use of electric motors had been pioneered, notably in Blackpool, and Bradford was the first city to switch entirely from horse-drawn to electric-powered trolleys in 1892.

Though the problem of supply had a temporary solution during World War I (because wartime industry needed heavy electricity production to function), after 1918 the issue rose anew, as only 6 percent of all households in Great Britain actually had electrical outlets. By 1920, 8 percent of all electricity was sold to households. By World War II, it was 27 percent, and by 1963, 41 percent, which is what engineers had hoped to achieve in 1908 (Forty 190).

Of all the reasons for limited electricity use, cost was the primary one; electricity remained expensive in relation to other methods of heating and cooking, partly because such little of it was needed (few households had more than two or three electrical outlets aside from the lights). It was also not efficient, especially where cookers were concerned. Stoves took a long time to heat, and the temperature control was hard to estimate. Time and again consumers were told that electricity would soon dominate the kitchen and prove safer than gas outlets. Gas companies, however, were able to rely on a proven track record to offer amenities in the present, thus suggesting they understood better the needs of the housewife (Forty 193). In addition, electric stoves did not offer any radically new or attractive design, which meant well-to-do families were unlikely to invest in the chic factor of a new appliance. This element, though seemingly trivial, is important because electricity since its appearance in the late nineteenth century had been labeled the clean energy of the future that would lighten workloads. At best it lengthened the day, but was that to do extra work?

The irony becomes clear in the numbers of advertisements and expositions that featured the bright new world of electricity. From transport vehicles to lighting to entertainment, electricity provided it all, or would soon. For now, its cost was such (and it remains high) that generations of Europeans grew up in the twentieth century with an ambivalent relationship to electricity.

Once adopted, however, energy also acquired its own identity on account of cost. In France, the consumption of electricity until the early 1960s was extremely limited on account of the French government's limited distribution of power. Put another way, over 50 percent of all households could not plug in an electric iron: It would have shorted out everything else in the house. Then, however, a new industrial policy was put in place that included boosting the current available to consumers. The problem was convincing them of the new availability of power. To symbolize all this, the French electricity company (EDF) began marketing blue-colored meters. In fact, these were not needed (the old black ones could take the increased current coming into houses), but the point was to emphasize a new era in which modern meter boxes were heralds of modern consumption. The marketing worked too well. While average consumption rose 10 percent per year over the first five years of the blue meter, EDF ran out of blue meters. Its sales representatives ended up applying blue stickers to old black boxes certifying these were up to the new standards (Akrich and Méadel 36). The meters, however, ended up fulfilling a new function. As power users developed a stronger awareness of costs and when rates were higher, many thus learned to make more efficient use of their electrical goods during periods of lower consumption.

The quest for electricity that would be both cheap and in quantity included a variety of solutions from river dams to solar power. Europe has generally run out of ideal spaces to construct powerful electrical dams. Solar power, like in the United States, is attractive, but expensive. For decades, the alternative was to construct nuclear power plants.

NUCLEAR ENERGY

Studies of the atom and its applications before World War II prompted the popular press in the Western world to publish affirmations of a future where entire cities would be lit up thanks to nuclear power alone. With World War II, however, the impetus came from the military side and led to the development, testing, and use of the atomic bomb in 1945.

A Series of Uranium Bars

Nuclear power after World War II thus became a symbol of humanity's ultimate self-destruction to some, but also of its savior when applied peacefully. In most of Europe, nuclear energy planning became the ultimate symbol of technological achievement and of growth (Hecht; Kupper 137). Several nations set up a program that in theory should have become the central element of a continent self-sufficient in producing, and even exporting, electricity. Several factors, including government subsidies, the lack of alternative sources of energy, and the oil shocks of 1973 and 1979, encouraged further the development of nuclear power. Nuclear accidents, however, combined with ecological awareness, slowed it down.

There exist two ways to create nuclear energy: fusion (the combining of two atoms' nuclei) and fission (the splitting of an atomic nucleus into two of roughly equal mass). Currently, fission is the only one used in a stable manner on an industrial scale.

Today, there are 448 nuclear power plants in service in the world. In continental Europe, some nations made the decision to rely almost exclusively on nuclear power plants. In France, 58 of them supply between 80 and 90 percent of the nation's energy and are used to export electricity to neighboring countries.

Taming nuclear power for civilian purposes became a central task for many governments the world over. And the faith in the technology was considerable, despite its documented destructive power. The concept of the atomic age illustrates the dominant position rapidly acquired by this scientifically based production of energy as the technology of the future.

The production of electricity on a relatively large scale occurred in the middle of the 1950s in the Soviet Union and Great Britain, and in the 1960s,

The Bugey nuclear power plant near Lyon, In Eastern France. Ecologists and nearby residents have long decried the impact on the landscape that such massive constructions caused (on this picture, the plant proves bigger than the village in the forefront). However, access to affordable electricity has been the selling point for many plant owners as they negotiate with communities for the right to build. Source: Courtesy Région urbaine de Lyon.

nuclear power began to become a competitive alternative to fossil fuels. In other countries, the technology became a political issue.

In Sweden, nuclear power became a political issue in the early 1970s, about the same time that the first Swedish commercial reactor was commissioned. Politically, parties capitalized on this, as center and left-wing groups drew most nuclear opponents. The Three Mile Island accident in Pennsylvania prompted the Swedish Parliament to initiate a referendum in March 1980. The left parties garnered the most support, and the parliament decided that 12 reactors would be permitted for their technical life. Decommissioning would occur "with regard to the need for electricity and the maintenance of employment and the common good" and would be completed by 2010 (Wormbs and Jernelöv 28).

The Chernobyl accident did not result in any technical changes in Swedish nuclear power stations, which were of a different type. But politically it meant that the Swedish Parliament in 1988 made the decision to begin decommissioning power plants sooner.

The decision was overturned in 1991 after a parliamentary compromise between parties. Decommissioning would only begin once acceptable, inexpensive alternatives became available. In 1997, the law on decommissioning nuclear power was passed, and the first power station was closed in 1999. The Swedish state is also negotiating with the power industry on decommissioning. The technical life span has now raised a new problem. Projected to be 25 years, a typical power plant may in fact last up to four decades. Consequently, decommissioning Swedish nuclear power faces a conundrum: close down an already paid for industry that could still pay off dividends, or go the safety route and invest billions in new energy production infrastructures. Furthermore, recent polls suggest that the public, though affected by Chernobyl, favors by a clear majority some form of nuclear power due to the lack of alternatives.

A notable exception to the case for nuclear shutdown involves the French state, which provides up to 80 percent of all electricity consumed in France. Here the technology has a unique position in Europe. In the 1960s, for example, nuclear energy seemed to offer the definite solution to all energy problems, as one nuclear power plant could easily replace several hydraulic ones, which were harder to implant. Whereas the French government, through its monopolized electrical company, was able to start planning at will the establishment of power plants throughout its territory, planned installations began to face far stronger opposition. The popular view is that environmental movements alone have forced a slowdown of nuclear power. As the Swedish case showed, politics play a big role, too. The same goes for Switzerland.

Swiss power plants have faced a variety of opposition to their development, though there is recognition that alternative sources of energy are not forthcoming yet. One saga, however, suggests that bureaucratic inertia as much as environmental and local concerns plays a role in the success of nuclear power.

In the case of a planned nuclear power plant at Kaiseraugst in Switzerland, the early phase boded well for the private operator eager to build it. In 1966, investments in nuclear energy production were welcome, and the opposition companies faced was primarily from the competition (Kupper 292). The winning company, Kraftwerk AG, was in fact a consortium of several companies whose membership numbers kept shifting as it sought more support to spread the investment risk. By the time the site had been approved and the company was in the process of gaining all associated authorizations, almost a decade had passed. The delay in obtaining clearances went back to the fact that not only were local authorities to give permission to build but so were cantonal and federal authorities. The initial materials to be used in construction turned out to be insufficiently strong, and the proposal to flush the used waters into the nearby Rhine River (which then flows north into western Germany and up to the North Sea) raised early ecological worries. An alternative solution, to build cooling towers, collided with aesthetic concerns about the regional landscape. This time-consuming process took the project into the 1970s, by which time a clear antinuclear attitude had begun to form. Interestingly, the company, which had first offered complete openness, failed to account for this new public mood, assuming official consent alone would matter. In so doing, it exhibited a kind of double weakness, as a historian of the project has shown: Not only did it underestimate the gathering ecological momentum, but its openness did not translate into flexibility in the planning of the project (Kupper 292–95).

Things boiled over in April 1975, when project opponents took over the construction grounds and occupied them until June, thereby occasioning national coverage and discussion about the benefits and dangers of nuclear power plants. When attempts at dialog were restarted, multiple actors were involved: Civil authorities argued for the absolute necessity of the project for energy supply reasons; the company argued for proceeding soon (it was not making any money on a blueprint power plant); and the ecological movement, drawing its strength from various kinds of opposition (local residents, militant greens, social activists), demanded cessation of activities. Each side had hardened its position so much that dialog became synonymous with confrontation (Kupper 150–90). Though the 1979 TMI accident only comforted respective positions (with the company arguing such problems could not happen at a modern power plant), the Chernobyl accident in 1986 was the nail that sealed Kaiseraugst's coffin. Kraftwerk AG announced its withdrawal from the project. In effect, lack of communication as well as weakness on the part of the operating company had begun sinking the project. Ecological concerns finished it off. In many ways, though, this had less to do with the actual dangers of nuclear energy (such awareness came about toward the end of the project) than with public concern that advanced technology was spinning out of control, regardless of its benefits. In 1988, without opposition, federal and local authorities approved the termination of the Kaiseraugst project and reimbursement of the costs Kraftwerk AG had incurred.

GOING GREEN

Much as nuclear power became the focal point of environmental reform, the general call in Europe for alternate forms of energy was based on broader environmental and social concerns. In the case of Germany, the Green Party over two decades went from a loose amalgam of concerned citizens to a full-fledged political party that included a foreign minister and vice chancellor of the Federal Republic.

In the late 1960s, Germany continued to experience strong economic growth initiated 15 years earlier. The political left wing, especially the Social Democrats (SPD), viewed this as the result of conservative policies that favored capital, but it believed that it, too, could benefit from all this progress. However, some followers of the left felt it was also the SPD's fault for ignoring the social impact of building low-income housing without attention to environmental issues: In some cases, there was as much noise in these satellite cities as there was downtown in the morning, yet deadly silence on weekends (see chapter 2). Thus, from the optimism of the 1960s, when the left thought it could change things and make a difference thanks in part to economic success, many left wingers turned away from progress to look at quality, not quantity, in the 1970s. Did a cubicle apartment and a TV suffice if you never saw your kids due to long commutes? Why live in the countryside if all you breathed were the fumes from the factory next door?

Basically, the Green movement came to stress that humanity and progress alone were not enough. You needed the surroundings to make sense out of your place in the world. The planet was an ecological whole. Socially, this meant all kinds of new ideas had to be considered. Some came across as perfectly valid (1 million parts of carbon dioxide on your highway to work is bad for you). Others were a little more esoteric and reflected the popular cultural mood of the hippies. The latter term was inoffensive, but under the heading of new age, it became a pejorative attribute for grouping anyone who claimed a need for going back to nature. Nevertheless, from the time of the first Earth Day (April 21, 1970) until the formation of the German Green Party in 1980, a loose coalition of themes and ideas began to take shape. The general motto was "we are neither left nor right, we are in front."

To articulate a federal program that would attract enough votes, however, was very difficult in light of the varied kind of followers the Greens attracted. They identified four pillars of activity. The first was ecological (saving the planet and its ecosystem from industrialization and unchecked growth). The other three involved social and political issues under the banner of pacifism.

In the late 1970s, people started joining the Greens in part in reaction to SPD Chancellor Helmut Schmidt's pragmatic rule and the decision in 1979 by NATO to install Pershing II nuclear missiles by 1983 in Germany. The result was that, in 1982, the Greens gained seats in the Bundestag, the West German Parliament. Though they played a role as an opposition party, the need to serve all constituencies, not just the ones concerned about nuclear power, proved difficult. Consequently, a serious conflict over the party agenda removed the Greens in 1990 from legislative federal power. By 1998, they came back into power as part of a Green-Socialist coalition that was more realistically oriented in what it felt was achievable.

Overall, then, while the Greens did not achieve any general goal of reform in society (partly because their program amounted truly to so-called ecotopia), they did achieve specific ends. Many of the pollution limitations, for example, were implemented through Green lobbying. Germany itself committed in 2002 to abandoning nuclear energy, though it has to rely on French nuclear-produced electricity to balance Germany's shortfall.

AFTERMATH

Two nuclear accidents raised concerns about the safety of nuclear power in Europe. The Three Mile Island event in 1979 was the first to introduce the need for clearer safety standards, but it was the Chernobyl nuclear accident that cast nuclear power as clearly dangerous in the public mind.

At Three Mile Island in Pennsylvania, parts of the nuclear core melted and radioactivity leaked out. The risk of radiation damage was assessed as slight, but reactions were strong, and in the United States, no further nuclear power stations were ordered after the accident.

The second accident occurred in April 1986 when a reactor at Chernobyl in present-day Ukraine exploded. Radioactive materials spread with the winds and rain to large parts of Europe, and in certain parts of Sweden, very high levels of radioactivity were registered. Recently, researchers have been able statistically to establish that the number of cancer cases in Sweden has increased as a result of Chernobyl. Much earlier, the number of dead and injured in and around Chernobyl had reached terrifying numbers. Chernobyl, especially, has become synonymous with the risks associated with nuclear power operation.

However, despite ecological pressures, many countries continue to operate their power plants for lack of alternatives. When they do interrupt operations, it is due to safety and cost concerns within the nuclear power industry rather than external pressures. This happened in 1998 at Crey Malville, France, and in 2006 in Sweden.

The risks associated with nuclear power were known early on, in light of medical studies done at Hiroshima and Nagasaki. Consequently, most European nations signed the Paris Agreement in July 1960 intended to legislate damages to the victims of nuclear accidents. Two fundamental rules apply there: The operator of the nuclear plant is solely responsible (not the constructor); second, victims have up to 10 years to link the nuclear accident to their health problems. Though seemingly simple, the question of damages is far from clear. In France, where there are the most nuclear plants in Europe, victims would get the least amount of damages under the law, whereas Germans would get 12 times more. This legislation has yet to be revised in light of the European Union. The United States did not sign this agreement, as such incidents fall under the Price Anderson Act.

The original nuclear age has ended, but its specter remains. For example, two of the four 290-foot-high Calder Hall towers at Sellafield in Cumbria were brought down in September 2007. The power station was opened by the queen in October 1956. In its early life, it was primarily used to produce weapons-grade plutonium, but from 1964, its main role was providing commercial electricity. The reactor was shut down in 2003, but nuclear waste is still reprocessed at Sellafield. Nuclear energy, then, is central to the needs of many Europeans, but it is far from nonpolluting and may end up costing more in the long run.

9

TECHNOLOGY IN THE HOME AND THE OFFICE

The penetration of technological machinery into the private sphere was a slow process. Energy itself was not trusted for a long time. Whereas streets, for example, saw electrical light installed by the late nineteenth century, many households did not have outlets until the 1930s. And just as the car displaced the horse within two decades, many standard technical features in the kitchen and the living room did not come until after World War II in European households. In fact, the "domestic saturation" of household gadgets and tools did not occur until the 1970s. The home, often identified as the private sphere beginning in the nineteenth century, was cast in gender terms and became, according to middle-class ideals, the safe haven where the man of the house could return and enjoy time with his family. Managing the place in preparation for his presence was his wife, leading an army of domestics. This cheap labor was another direct impact of the Industrial Revolution, and servants were expected to do everything from cooking to cleaning, dusting, laundering, fetching wood, and caring for the children.

Training courses in fields ranging from sewing to typing and even telephone operating changed this state of affairs. Initiated by social-minded reformers, such courses were intended to help working-class women rise to a level where they might eventually become part of the middle class. While this hardly happened, the impact was a dearth of domestic help, which World War I would accelerate as women moved to factories to take on the work their male counterparts at the front could no longer do.

GAS

Gas and electricity producers experienced challenges to household consumption as they promoted their services. In the case of gas, the need to find other uses than simply lighting prompted some companies to offer consumers special incentives. Some tried daylight fees that would be cheaper, but generally, most producers adopted the tradition of installation and connection at a very low cost. This practice extended later to electrical and phone connections. In Paris, such incentives saw a doubling of gas oven rentals over a decade; by 1900, over 300,000 households had one. To further convince the clientele, many companies opened stores with display models that explained how easy gas was to use. In the case of gas ovens, salespeople recorded the enthusiasm of women who were used to lighting a fire with wood and coal and waiting for the oven to warm up. Gas in fact was a break from the past in terms of efficiency, but the notion, to quote a commentator, that it emancipated women by removing the need for them to spend the day by the oven was false. It simply allowed women at home to focus on other domestic activities. What these anecdotes show is that people adopting a specific energy source did not always do so on account of what that energy could do but on the basis of other tools and machines they used.

The application of electricity in the home, first through lighting then through other contraptions, became the primary sign of modernization for many families. However, electricity was feared as much as it was sought, and manufacturers had to work extremely hard to convince their clientele of the advantages of going with electric current. Indeed, fearing a new technology was only part of the equation; clinging to tradition was another. Many electrical appliances were to be used in the kitchen, and marketing departments everywhere sought to advance the agenda for more automatization and fewer servants: the middle-class ideal was to be modern, not traditional, or so the advertisements suggested (Frost, "Machine Liberation" 114).

Just as bathrooms became more sophisticated and clean, kitchens grew bigger and more rational, at least in the eyes of designers. This new domesticity, where the wife/mother would direct servants in a newly refurbished kitchen, was a further evolution of the private sphere that had first crystallized in the nineteenth century. Partisans of modernism suggested that science and technology applied just as much to the woman in the home as to the man at work. One (male) scientist even suggested that while he wished not to limit educated women to the kitchen, redefining that area as a laboratory would help her reclaim an honorable identity in a technical world (Frost, "Machine Liberation" 119–120). An imaginary dialog from an advertisement reflected the ideal:

> Guest woman: "My dear Friend, your house looks wonderful and so modern! Everywhere there is electricity."

Hostess: "Of course, I keep up with the times as they say . . ."
Guest: "And no troubles?"
Hostess: "Never any; I use only branded machines . . ." (Frost, "Machine Liberation" 125)

To be modern was not just to use machinery but to consume it. Design mattered and was reflected in the American-inspired models shown at the Domestic products show in Paris that became available to a small elite. When examining them, a journalist felt a laboratory, or even an operating room, had now entered the household: "With current ovens, one knows the cooking time of a roast or any other dish the way a patient's doctor knows his condition with a thermometer. Temperature is dosed the way

Gas companies would seek to offer many incentives to encourage consumers to shift to gas cooking. Though the advantages of gas over wood stoves are substantial, it did not cut down on workloads in the kitchen. Here, an extremely versatile homemaker appears to be baking for a regiment. Or perhaps they are teenagers? Source: 1925 advertising postcard, author's collection.

a chemist measures dosage. Different heating areas allow for the preparation of multiple dishes. The culinary arts gain in practicality without losing in sophistication" (Lenief 106).

THE SEWING MACHINE

Sewing machines, developed first in the United States, soon expanded the market of their inventors overseas. The Singer company in particular, which had earned its first patent in 1851, quickly attracted European entrepreneurs, following its first exhibition of sewing machines at the Crystal Palace exhibit in England. An attempt to first license their patent on the French market met with failure due to the debts the licensee incurred as well as the outbreak of the American Civil War (Davies 300–304).

Josef Wertheim, for example, learned methods of mass production during his time in the United States but returned to Germany where, in 1868, he founded his own sewing machine company intended to compete with the massively successful American Singer. By 1883, his 600 employees were turning out the 35,000 of his highly successful Electra. Like the Singer, though essentially an industrial tool, the Electra also had black doilies and golden highlights so that it would pass into a domestic setting.

Singer was not left out in the cold, however. Its German representative George Neidlinger covered much of northern and south central Europe. Neidlinger moved to train representatives to meet with customers but also to return and collect the regular payments (most machines were sold on a credit basis). However, such practices sometimes backfired, as agents moved to sell to higher risk customers and found themselves later unable to collect the promised fund. Nonetheless, with experience, the company began raking in sales and became a heavyweight competitor for local manufacturers. The setting up of factories in Europe (Scotland and Austria, both chosen for economic reasons) also made a difference by shortening the time between orders and deliveries.

Salesmen, it turns out, made the biggest difference, not the product itself. To ensure that none of their traveling canvassers sold in a way that would hurt the reputation of Singer's English operation, managers offered commissions but also made them responsible for evaluating the credit and character of each customer (Davies 309). Further incentives such as an insurance fund were also offered to ensure increased loyalty to the company.

The success of the sewing machine is best explained through the tradition of women's work in Europe. Though dominated by tailor guilds (associations that decided who may exercise a trade) since the Middle Ages, the practice of weaving and cloth making within the household had been part of most rural (and later urban) families' tradition and was an expected source of income. Time and again, however, the practice of women sewing garments and clothes had been challenged, as it was deemed a threat to established men's work in cities (Quataert 1124–26). This state of affairs changed as industrialization brought in the factory and did away with many guilds. Governments, however, stepped in to protect what were deemed women's traditional roles. Romanticized in the eighteenth century, the role of housewife was hotly debated among lawmakers. In the German state of Saxony, for example, working in the private sphere of the household was viewed as a noble but restricted activity, and legislation appeared that allowed women to continue to work as seamstresses. By the 1870s, a solid garment industry continued to exist in the city of Berlin and relied heavily on individual workers in the home (Quataert 1138–41). This was due in part to the fact that all girls in primary schools now had to take

The sewing machine was an ideal tool for women's housework, as it allowed for an efficient variation of the traditional putting-out system (aka cottage industry). It also became an attribute of the middle-class woman. This romantic greeting card from 1905 suggests the young woman longs for her companion to return home soon and share in the delights of the safe haven she is expected to manage for him. Source: Author's collection.

needlepoint and other forms of sewing so that they would all have a marketable skill. Because factories began setting sewing machines to accelerate garment production while still employing female members of the working class to keep costs down, the sewing machine itself, though a substantial investment, began to appear in working-class homes as a means of subsistence. In middle-class homes, of course, the element of survival simply did not exist, but a high-end sewing machine did reflect status. Nonetheless, the sewing machine did not bring about gender equality of any kind. Male tailors who underwent a professional training of up to three years, for example, could apply for a pension once they retired. Women in a similar situation in Germany were denied the support because their work, though similar, was deemed housework (Quataert 1146).

Women's work and salesmanship are key factors in explaining the widespread success of sewing technology in the home, but advertising also played a substantial role. In France, for example, where a similarly poor working class became a major sales target, advertising brochures included invitations to view machines in luxurious stores and apply for a credit line that would enable the acquisition of an onerous machine. Indeed, the machine was no longer simply a means of subsistence. Its placement at the center of a palace-like department store gave people the illusion of a middle-class existence (Coffin 754). Credit payment was risky (seamstress work was usually a seasonal occupation), but again, the notion that one could be modern, and therefore competitive, often proved too hard to resist. Indeed, the advertisements often combined very feminine figures mastering a new technology. This had not always been the case. Early advertisements featured men, but by 1900, women alone appeared in such advertisements. To sew, even with a machine, was now deemed women's work. Machines came with elaborate wrought iron bases and suggested they would be subjects of conversation during ladies' visits. The actual functions (sewing holes and other domestic chores) were not mentioned (Coffin 763). This was all the more the case by the turn of the century, when the "new woman" image appeared.

The new woman was a woman characterized by intelligence, strength, and sexual desire, in every way man's equal. The new woman was determined to take charge of her life, and she violated almost every social code inherited from the nineteenth-century separation of spheres that consigned her to the stereotypical position in Victorian society of the angel at the hearth. Advertisements targeting any woman attracted to the new ideal typically showed a female figure in control of the sewing machine much in the manner one would control a locomotive (Coffin 767). Sewing machines were often shown side by side in department stores with bicycles, another tool of independence, and also produced with wrought iron.

Not everyone agreed with the value of the sewing machine for women. Many physicians were convinced that such technical challenges were too much for a women's brain and were the cause of stress and exhaustion and even sexual arousal. The evidence for any of these assertions was scant and represented more of a male misunderstanding (and fear of mismanaging) of women's health, but such notions continued to exist until World War I. By then, the sewing machine had become a central element of female domesticity, one that later on would appear on any wedding wish list.

TYPEWRITERS AND THEIR OFFSPRING

Though there were several attempts at inventing typewriters, early endeavors in Europe (notably Italy) failed to reduce any public interest in the nineteenth century. On the other hand, in the United States, Christopher Latham Sholes's patent eventually translated into a successful machine, improved upon by several manufacturers. However, skepticism remained strong in many European nations as to the value of typewriters.

Simply put, the act of writing in long hand was as valued as a form of human contact, and the depersonalization block print brought also arose suspicion. Well into the late twentieth century, while employers accepted typed resumes, the letter of application was expected to be handwritten.

In this context, the success of indigenous manufacturers is all the more interesting. In 1908, Camillo Olivetti founded his own typewriter company and, within three years, was challenging American competition successfully and expanding into other European countries. By the early 1930s, it sought to introduce the first portable typewriters. What attracted the public, though, was the design of the Italian machines. Whereas early typewriters had an image of bulkiness and even discomfort, Olivetti had taken to reshaping them: By smoothing the contours and changing the color (all typewriters used to be black; Olivetti's was grey), he gave consumers a combined sense of lightness, functionality, and ease of use. The typewriter became less scary to use (Kicherer 25).

Olivetti was a pioneer, but others followed. Many offices used duplicating machines to print internal memos and lists. Their design suggested a miniature industrial steam-powered wheel. A British company, Gestetner, was the first to seize on the notion that office tools should not look like outcasts of the first Industrial Revolution. Calling on a French-born American designer, Raymond Loewy, Gestetner was able to market a duplicator with a completely enclosed mechanism well into the 1950s. The "face-lift," as its designer called it (Loewy 82–84), also suggested something important about the operators: They used the machine but had little interest in its internal workings. Usable technology did not require knowing its engineering.

The typewriter—and the profession of typist—came to represent a goal in the United States for many women starting to work at the turn of the century, but it was also a hindrance (Gardey 319). On the one hand, typing came to be seen as an acceptable profession for women, when standard middle-class ideals would have expected they remain home and manage the safe haven while their husbands engaged the business world. Socioeconomic reality required that young women and even widows join the

N° 8 — Bureaux

Typewriters in offices before World War I. Though men in the late nineteenth century were associated with secretarial work, the typewriter contributed to replacing them with cheaper labor, usually unmarried women. Source: Anonymous postcard, author's collection.

workforce. This association of woman and machine was partly the result of marketing efforts. In the United States, the Remington company, one of the first typewriter manufacturers, used women for its demonstrations but preferred to hire female workers with an educated background. The association of the piano with the typewriter came quickly and was also noticeable in France at the turn of the century (Gardey 326). In fact, women thus gained an unlikely entry into the office world. By the 1920s, however, said entry came with restrictive dimensions.

In parallel to their newfound freedom, women also came to earn less than their male counterparts. The French administration, for example, moved to hire women alone. Since their function was limited to typing, they worked three times as fast, for a third of the men's salary.

Changes also occurred *around* the physical location of typewriters. Well-desks, where the typewriter was placed at a different level from the rest of the desk surface, already existed in the nineteenth century, but in the 1920s, adjustable surfaces were added. Furthermore, as American-based scientific management took over offices in Western Europe, secretaries' desks generally shrank in size. Since women's functions were limited to typing, it appeared as if it would be superfluous to waste space. Instead, typical offices of the interwar years came to feature several desks with typewriters. Reminiscent of a factory floor, office layouts were distinguished mostly by the fact that occupants wore city attire rather than factory overalls (Forty 133).

The typewriter as an object of fantasy in the interwar
years. The mood this advertisement seeks to convey
suggests that the secretary is attracted to the machine
itself, perhaps for its ease of use. The advertisement
also assumes that women alone will be typing, but
it is directed at men (who purchase the office equipment) by making the woman young and attractive.
Source: Advertising postcard, author's collection.

Last but not least, secretaries, aside from taking shorthand, were now
expected to type with the use of a Dictaphone. This way, the typist would
stay at her task while the executive could dictate what he wanted at any
time. However, interwar years dictating machines, though marketed in
Europe, were unpopular. Not only did they displace secretaries who knew
shorthand (and welcomed the diversion to move from their desk), but the
contraptions were bulky and the wax cylinders onto which voice was recorded did not restitute voice well, thus slowing rather than accelerating
the pace of work.

It was not until the 1950s that magnetic tape dictating machines became
standard. Their redesign and eventual miniaturization made them easy
to carry, though many executives balked at bringing them home. Indeed,

though advertisements suggested the ease of use made it child's play in the private sphere (Forty 140), this suggested that the home was no longer separate from the office, and bringing work into the family realm might not only be practical, but even expected. In the meantime, the European home was undergoing a slow but irreversible transformation.

PRIVATE-SPHERE IDEALS

The home underwent its transformation into a private sphere following the French Revolution and the Industrial Revolution. Middle-class projections of wealth included the introduction of the parlor area, essentially a receiving lounge separate from the place where the family might congregate more informally. The set up of the home involved a new domestic architecture but also improvements in technological know-how as pertained to furniture making.

Because of a quest for comfort associated with the opulent, if severe, Victorian style, a major innovation in furniture involved the spring. It either replaced or supplemented horsehair and feather stuffing in chairs and beds. Englishman Samuel Pratt's patent on the spiral spring mattress, for example, became a major consumerist longing. Interestingly, the ideals of luxury clashed with demand, and the furniture trade in Europe was slow to adopt mass-production practices. It would take World War I to complete the change.

Because the airplane became a war technology in 1914, yet relied on wood (the change to metal did not come until the late 1930s), the drying up of all furniture supplies followed the mobilization of the troops. Not only were workers shifted either to war factories or the front, but wood became scarce. Furniture making became aircraft making.

After the war, industries that had for the first time been introduced to scientific management in the war often attempted to revert back to their prewar traditions, but the revolution in efficiency was only slowed down, as suggested by this enthusiastic editorial commentary:

> Having regard to the world shortage of furniture, perhaps the most valuable result of wartime experience will be the education of our manfacturers in mass production. The absolute accuracy and precision demanded by the A.I.D. [Aircraft Inspection Department] for all airplane work accustomed heads of factories to the idea that intensive production need not necessarily mean bad work. (quoted in Edwards 74)

Furniture making, then, was considered a fashion industry rather than a technological one and generally expressed misgivings about the training of workers into single tasks: What was needed was someone versed in all aspects of the craft, even if training took longer. Though some companies did adopt a more streamlined approach to furniture manufacturing, they also found that the introduction of new materials, such as aluminum, into civilian life, was not as easy as it appeared.

Different kinds of economic necessities affected the use of aluminum in everyday life. Nowadays it is taken for granted, most notably as a component in soda cans. Prior to World War I, however, it had trouble finding a usage in everyday life (some manufacturers even tried selling it in the form of postcards). By 1919, a few companies tried to incorporate aluminum into furniture making (notably in England) but found this could be done only at the level of chairs' structure, for fear of turning away clients used to noble wood. Anecdotal though it appears, this episode also reflects a wish on the part of those who experienced the war to return to a pre-1914 life where technology did not appear to pervade every aspect in life. Hand-made furniture would continue to exist, but the march was on to adopt mass production. The iron tubing, as expressed in Marcel Breuer's introduction of the cantilevered metal chair, offers a case in point. The home would eventually receive modernist material, sometimes because it was trendy, more generally because it was cheap. Other transformations would take longer and be part of a reform of social mores.

CLEANLINESS

The advent of running water in the home encouraged not only a new application of the principles of cleanliness that nineteenth-century medicine had begun to stress but also a recasting of the bathroom (see chapter 10). Since the nineteenth century, the family home had become the new safe haven to the social unit of the nation. Victorian England had crystallized such a notion, though it was to be found in all industrial nations. As such, the outer business world was dirty, and it followed that the inner sanctum should be spotless.

It should be emphasized, however, that this was a middle-class urban ideal. Grooming overall remained limited outside the city, for access to water was limited in the countryside (and would remain so until after World War II). Furthermore, an old belief still followed claimed that to bathe made one's skin soft and therefore vulnerable. Dirt implied health. One might splash water on the face and hands prior to a religious service or a special event, but that was all (Prost 85).

As middle-class ideals became the new standards for ideal living conditions (thus displacing the nobility), the notion of cleanliness may have become accentuated as a means of distinguishing old and new but also social background. Cleanliness became a factor that helped justify power relationships between castes.

Bathrooms reflected this changing state of affairs. First installed with running water in the late nineteenth century, they gained in engineering sophistication rapidly. Designers of sanitary ware stressed white and chromium as defining traits of the proper modern bathroom. Physical cleanliness and efficiency were part of aspiring to become modern (Forty 117).

> There is no room where cleanliness and neatness are more necessary. . . . No bath should be fixed [have cabinet work around it] except in such a way that every part, underneath or at the sides, can be easily got at and cleaned. Woodwork, whether as a rim to the bath or as a casing around it, is to be avoided at all costs. (quoted in Forty 167)

This excerpt from a British "household encyclopedia" of 1923 suggests the paradox in all its splendor: "The place where one entered dirty had to be spotless. The hygienic argument, however, quickly gave way to a social one, whereby the housewife was a guardian who fought dirt for the good of the family" (Cowan 22).

Following World War I, a building boom allowed middle-class families to move into housing that had all the amenities. Newly constructed workers housing also came equipped with bathtubs and sinks, but it seems that access to such amenities was not always understood. Jokes circulated among upper classes that bathtubs were used in blue-collar areas to store coal or even raise a rooster or two. Apocryphal though the rumor may be, it does point to the fact that social change within the private sphere remained a very slow process.

The reality was perhaps more difficult to grasp. Many in the working class aspired to live in comfortable surroundings or at least to modernize their current residence. Because water supplies, when available, often came through the kitchen (with restrooms limited to a single unit for several apartments), enterprising renters and owners had bathtubs installed in the kitchen, with a bench cover as a space-saving measure (Lenief).

NEW TOOLS IN THE WAR ON DIRT

The housewife, then, had weapons at her disposal. One was the vacuum cleaner. Also invented in the nineteenth century in the United States, the vacuum cleaner did not see efficient commercialization until the 1890s. By then, notions of better hygiene (especially removal of heavy dust) combined with those of scientific management (a concept first developed by American Frederic Taylor to make factory work more efficient) suggested vacuuming was the solution to several problems. Not only would dirt be removed from the private, clean sphere, but cleaning would be done more efficiently (Gideon 549).

Initially, private companies traveled to houses and buildings and vacuumed out the dust into containers. The engine was often set up on a carriage. This service, however, assumed high wealth on the part of private contractors. By World War I, the basic principle of the vacuum cleaner had been identified, and standard models appeared on the market. In fact, these early contraptions proved not to save time but to only be more efficient cleaners: The claim of time savings thus disappeared from advertisements (Forty 176–77).

The cost of vacuums and the need for electrification of one's house or building resulted in the limited use of such contraptions in the interwar years, but the general public was very much aware of them. As one journalist put it when commenting facetiously on the Arts Ménagers show in February 1931,

> Dust, once running wild amidst broom strokes and simply trading spaces is now mastered and bagged. Vacuum cleaners won't let it be. They chase it into corners and seek its destruction. (Lenief 107)

Irons evolved from the Middle Ages onward as tools favored by tailor guilds and later households. Though initially consisting simply of flat stones or metal pieces heated up in the fire, the design quickly evolved in the nineteenth century in response to the expectations of proper private sphere standards. For example, irons often had ash deposits from heating near the fire, which ruined white clothing. Newer models opened up to receive hot coal. With the appearance of gas lighting and heating in well-to-do households, however, new models came to include gas-heated irons.

Electricity, of course, brought a new stage, but this was long in coming due to costs and practical issues. Irons remained extremely hot and ruined any fragile fabric that the operator might mistakenly overlook. Temperature regulators appeared in the 1920s, and 30 years later, a few steam models were introduced, mostly for professional laundry services. Another factor in the adoption of irons into the household was the lightening of the iron. Early tailor irons weighed up to 30 pounds, as the lack of constant heat and humidity was compensated through added weight that would flatten creases. By the middle of the twentieth century, irons were closer to 10 pounds in weight and quickly lost a few more. Though ironing was still considered a physically tiring task, it seemed more manageable.

Some manufacturers, inspired by American models, tried to modify factory machines (which used the principle of combined pressure and steam) into house steam presses that allowed the removal of the ironing board in favor of the press itself: The housewife could thus, or so the advertisements went, sit at the dining room table and simply close the ironing press with an easy lever, wait a few seconds, and shift the fabric to complete the task. Ingenious though it was, the design overlooked a combination of habit and practical issues. The characteristically small European household could easily store the iron and the ironing board. The footprint of the ironing press, on the other hand, assumed that at least half of the closet space's base would be used for storing it.

Although the nineteenth century saw the rising awareness of cleanliness for health and social reasons, it is the twentieth century that came to embody the actual application of such ideals. Sweeping, but also all manners of cleaning and laundry, were, of course, high on the list of proper household items to be managed. The advent of electricity and its introduction into the home gave new impetus to manufacturers of household tools to claim huge time savings.

This matter of time brings about a dichotomy. On the one hand, the ideals of private life as they were defined in the nineteenth century suggested the role model wife was solidly middle class and did not work, preferring instead to maintain an impeccable household for the husband to come home to. In so doing, however, she was expected to manage a cast of servants. On the other hand, the ideal of the housewife cast in a patriarchal system was precisely to *not* work. Housework, therefore, could not be considered work. At best, it could be a craft.

This did not stop entrepreneurs from attempting the construction of various washing machine models with constant improvements. The impact of World War I, with men at the front and many servants leaving to work in factories as part of the war effort, became an opportunity for advertisers and paragons of middle-class taste (such as the British *Ladies' Home Journal*) to argue that servants were a burden and that electrical contraptions would soon become the ideal servants: An initial investment would easily be recouped since savings were then incurred on clothing, boarding, and salary. The way consumers perceived the availability of vacuum cleaners, dish washers, and washing machines, however, was otherwise. In fact, servants remained part of British middle- and upper-class standards. The price of the machinery was such that, in fact, whoever could afford, say, a vacuum cleaner (often given female names by manufacturers to suggest a humanized machine), could afford a servant to operate it. Many housewives never used the vacuum, leaving its operation to house help (Forty 212–13).

Where households without servants were concerned, the illusion was that machines would indeed free housewives, or at least not make them "work." Key to this illusion, however, was proper styling. Initially, many household machines, be they vacuums or food mixers, looked like factory tools. It was not until the 1950s that design actually began to affect the look and attractiveness of machines in the home in response to consumer taste. The German company Braun, for example, pioneered the notion of a mixer that came across as light and discreet, even though the machinery the plastic and metal covers hid was just as solid (and bulky) as a more traditional design that might have been found in restaurant kitchens working to mix in quantity (Forty 220).

Early washing machines were derived from other contraptions, such as centrifugal extractors (they were modified to extract as much water as possible from wet clothing).

The British trade association, for example, published *The Lady, the Linen, and the Laundry,* which discussed the difficulty of identifying fabrics to wash them properly (Mohun 81). In this context, laundry services became legion. While it was common to have laundresses or hired help take the clothing to a nearby washing shack, the formal development of laundry services also appeared in cities. As of 1904, in England, about one-third of all laundries used power machinery (Mohun 49). In the latter case, formal floor plans and engineering advice were available. Men ran and operated these centers, with only a few women rising to managerial positions. Women in fact viewed such laundry services with suspicion, handing over their loads only when emergency arose. Otherwise, the very business of laundering, whether mechanical or by hand, still was viewed as part of the quintessential workload (and identity) of the housewife (Mohun 69). Though business did eventually gain a new clientele, the acceptance of the service was slow.

Some women, however, established a kind of cottage laundry industry by running operations out of their home or from a nearby rented space. Other married women would take up part-time work there. Though eventually displaced as hand washers, women remained active as ironers, for the ironing process was among the least mechanized (irons used to be heated on a special stove and had no temperature gauges) and, thus, the most skilled.

Apartment dwellers in particular made extensive use of laundry services, as accessing a washing area, let alone having one in the home, remained a difficult matter. This would change in the interwar years.

The 1920s and 1930s witnessed the appearance of the early washing machines and plug-in irons. Earlier prototypes had failed to attract any clientele. However, the adoption of such contraptions hinged on two factors: marketing and affordability. In the case of the former, consumers were often receptive to the notion of bringing the laundry back in the home, where it would be washed without other people's laundry. Technology was helping reassert the private sphere, supposedly cheaply and efficiently. While advertisements claimed that one could be done washing in at most an hour, women in the home reported spending much more time working on the contraption and keeping the laundry from becoming damaged. The impact of this shift, however, was also the slow demise of many laundries and the consolidation of businesses.

Adopting a new machine takes time. In many ways, it is more about the human operator adjusting to the presence of machinery than the difficulty of operating the contraption itself. Washing machines offer a perfect case in point. Advertisements of such machines in 1950s Great Britain suggested it would cut on time spent doing household chores (Forty 210). In reality, a survey of that era suggested domestic chores required as much, if not more, time at home. By the 1970s, said time had further increased. Two factors help explain this development: first, the reduced number of

VISIT OUR GLASGOW SHOWROOMS OR WRITE FOR OUR LATEST BROCHURE M.4

EZEE KITCHENS LIMITED
341a, SAUCHIEHALL STREET · GLASGOW, C.2
LONDON DISTRIBUTORS . . . . TEMPLE & CROOK LIMITED, 6-7 MOTCOMB STREET, S.W.I

"Fantasy will set you free." This British post–World War II advertisement suggests that the modern kitchen and its associated machinery will free this woman to leave her home and go about town. In reality, modern appliances created new expectations, including the need to clean the home more often. Women who were homebound were likely to remain so, especially as social mores still assigned them the care of children. Source: Astra, Geneva, Switzerland; reproduced by permission.

domestic helpers assigned to these tasks in a middle-class household; second, the increased expectation associated with hygiene, which expects more washing, in different loads, and with the consequent increase in ironing (Akrich & Méadel 35). The machine, it turns out, had indeed cut time, but leisure did not replace time gained. In fact, many women ended up spending as much or more time operating machinery.

CONSUMERISM AND TECHNOLOGY

One way to ensure people would buy new appliances was to exhibit them in the context of a novelty fair. In the tradition of the universal expositions of the nineteenth and early twentieth centuries and of early guild fairs, several yearly gatherings stood out in Europe and sought to draw consumers to admire and eventually acquire machines.

Style was notable in the industrial arts, but did it sell? In fact, the public had to be convinced of the need for new machinery in the city and therefore in the home. Inaugurated for the first time in 1923 and disappearing in 1983, the *Salon des arts ménagers* held almost yearly in Paris became the showcase of domestic mechanization (Frost, "Machine Liberation" 129). Though a great success in terms of attendance throughout the years, the intent of the organizers, namely, to create a society of consumers, failed to materialize. The items shown were often far too expensive, but like other shows cited here, they offered an escapist moment. Visitors might buy a lamp or even request a catalog from a particular brand, but overall they came to glimpse at what might be if their dream of opulence came true. One of the hindrances to making such dreams on display into a reality at home included the difficult economy in the 1930s, 1940s, and 1950s.

Opulence did not just involve radios or washing machines, however. The phonograph, for example, was also a measure of taste and opulence. When Thomas Edison invented the first working sound recording device in 1877 (he recorded himself reciting "Mary Had a Little Lamb"), he became the father of what one historian called "the industry of human happiness." Voice, of course, but also music, would soon be preserved for repeated listening. Others had preceded him in suggesting ideas related to recording sound, but his was the first functioning model, and it served as the basis for further improvements in Europe. In particular, Emile Berliner, from Philadelphia, helped set up a gramophone company that would manufacture records. The gramophone would be displaced by World War I. In 1898, in London, his company began selling a so-called dog model, which bore the famous logo "His Master's Voice" (Yarwood 160). It is in the 1920s that standardization of the recording industry made the record widespread. Its initial technology consisted of a single speed player, which spun the record at 78 rpm. Orchestras, but also poets and politicians, read texts. Radio shows were taped (or rather cut) onto records.

The post–World War II era saw new jumps in the improvement of recordings and players. The 33½ rpm format, made possible by better recording technology, was introduced in the United States and quickly spread to Europe. It allowed the storage of more information and became the standard LP (long play). The 45 rpm would come within a decade and be used to promote single songs to younger consumers. In parallel, the size of the players shrunk. Portable players would not be available until the transistor revolution of the 1960s (see chapter 6), but already a new challenge to the consumer appeared: sound.

With the concentration of urban populations into hastily built habitats after World War II, many a family would encounter the challenge of putting up with a neighbor's noise. The beat became grounds for considerable family tensions, long before the advent of earphones. A Punch cartoon reflects the notion well: a father seeking to please his daughter in a record player store agrees to buy one, yet he insists on getting a silencing box like the ones people could use in stores to listen to records before buying them. This consumerist trend, however, was slow in coming, as the economy stabilized only in the late 1950s in most countries.

In France, despite the economic growth that characterized the post–World War II era, observers noted a certain gloominess that failed to lift for over a decade and was accompanied by a distrust of technology in the home. War destruction, but also slowness in creating a basic support infrastructure, meant many areas were literally in the dark. Hardly 60 percent of rural households had an electrical outlet in 1946. Just as the traditional family returned to the fore, if only for a brief time, so too did traditional ways of doing things regain momentum. A 1954 survey showed that 7.5 percent of French families owned a refrigerator and 8.4 percent a washing machine (Maddison 88). Over the course of the following decade, however, a jolt in economic growth reignited consumer spending in home technologies, just as new products came on the market.

By the 1960s, aside from the items deemed essential to everyday life, such as refrigerators, washing machines, and of course radios and televisions (see chapter 6), Europeans were introduced to technological gadgets imported mostly from Japan and the United States. Sound machines designed to help sleep by mimicking waterfalls or rain came on the market in 1964 and cigarette cases with timers that released one cigarette every two hours to help smokers control their urges (a Swiss invention). Most such items would appear familiar to current-day infomercial audiences, and some would become standard fare. Western Electric's touch tone phone was considered a gadget, for no European phone line could use the feature, yet, the idea would eventually be adapted for business purposes within a decade.

Other materials reflected the increased purchasing power and leisure time available to Europeans. In the realm of photography, super-8 film cameras and projectors proved most attractive, but so did slide projectors. These competed with the increased availability (and consequent drop in prices) of color print film. While black-and-white photography had become an expected feature of one's vacations, it remained affordable mostly to middle-class families. Some of the top brands were already much sought after in the 1950s and 1960s, and enthusiasts used popular science magazines to inform themselves of the new through-the-lense advantages. Generally, cheaper cameras, mostly used ones, were purchased but remained complicated to use. The advent of the Polaroid and the commercialization of cheaper models in the 1960s and 1970s reflected a golden age in mass photography, but cheaper cameras like the 110 format and some standard 35 mm models limited Polaroid's expansion on the European market.

While one perceived investments in electrical home appliances as a need, the kind of necessity a radio or even a washing machine represented was another matter. Indeed, it seemed to be as much about peer pressure and the need to display social status as it did about making one's life more enjoyable. This state of affairs is perhaps best reflected in an anthropologist's summary of new acquisitions an Italian peasant made in the late 1960s:

Little by little they had inched ahead; the gas stove he had bought in 1956, the running water he had installed in 1963, and the Vespa [motorcycle] were all signs of this. It was a kind of catching up, he thought, catching up with the American way of life that was also becoming the Italian way of life. He saw it all on TV. (Pitkin 154).

CONCLUSION

The evolution of technological tools and gadgets in the home was the result of several factors, including the promise of a social utopia through the acceleration of everyday practices. In so doing, time availability was the central promise (and premise) and became the focal point of advertising campaigns throughout Europe. This did not change *who* operated most machines. In the tradition of the private sphere, most women were expected to continue maintaining the kitchen, do the laundry, and vacuum floors. Evolutionarily, European households did catch up to their American counterparts in terms of consumerism. Nowadays, new dishwashers, laundry machines, and stereo systems are also about design and consumption as much as function and effectiveness. Yet, the promised time gain gets spent outside the home on things that have little to do with leisure; while some activities are easier to perform, they are often replaced with other activities associated with the maintenance of a proper household.

IO

MEDICAL AWARENESS

Modern European history is closely associated with the history of medicine, as a series of new discoveries in the nineteenth century changed the way medicine had been practiced and understood for centuries. However, the medical field also reflects the rise of new advances in the United States, and the interchange, through cooperation but also through warfare, of new ideas for the treatment of patients. It is also in part due to the Industrial Revolution and its increased population concentrations that new practices in hospital organization and socialized healthcare, which included preventive medicine, appeared.

To survey all this here would be impossible. Thus, it becomes necessary to consider more simply the notion of medical awareness. Simply put, Europeans went from discarding medical precautions in the nineteenth century to incorporating medical awareness into everyday life. Diseases, such as cholera, the influenza pandemic, and AIDS will serve to illustrate how everyday life was changed and continues to change as a result of the achievements of the medical profession. However, we shall see that much depends on social mores, as in the case of procreation and abortion.

CHOLERA AND PERSONAL HYGIENE

In 1832, Paris was subjected to the first of several cholera epidemics that would affect Europe. Elsewhere in Europe, major cities experienced the same trouble. In London, over 600 people died, but some 14,000 became ill from an 1854 outbreak. The social disruption was substantial. Cholera can prove deadly by disrupting normal digestive function and dehydrating a patient in a matter of hours.

Nobody understood how it spread, however. Because of the stench associated with the diarrhea, many thought the disease was airborne. Quack prescriptions included mint boiled in water to "ward off the bad air," but generally people ran, sometimes to church to offer prayer. By 1832, however, there were noticeable changes in the reactions of the Parisian authorities. Because septic tanks did not exist yet, human feces and refuse usually ended up in a gutter in the street median. Piles might be collected on a weekly basis, but generally inhabitants were dependent on a strong rainfall to ensure the clearing of the center gutter. The authorities moved to close narrow streets where too much accumulation remained, and inhabitants began to collect money to have extra water brought in for cleaning. Chlorine was reportedly dumped into some public collection areas. The public understood instinctively the need for cleaning away the refuse, but the lack of microbial knowledge as well as limited access to restrooms still saw the water supply becoming contaminated.

The places spared from the cholera epidemics came to include the nicer parts of town, which could afford private cleanups independent of city budgets. The problem was not just about public services but also private hygiene. Bathing more than a few times a year still was rare, and it would not be until the mid- to late nineteenth century that a proper bathing area would be constructed into well-to-do homes. Barring that, public baths would have to do. Cities did move to initiate construction of such establishments, but convincing the public of the necessity to bathe was another matter.

The Western world as a whole viewed water with caution, as it was thought to bring disease. When the shift in attitude began to occur in the early 1800s, cleanliness also became a means to differentiate class. The working class came to be pictured not only as uneducated and wretched but also as unkempt. Thus, to be a proper citizen, one no longer simply exhibited morality but also cleanliness. To ensure that this notion would spread widely, philanthropic associations distributed brochures to explain the need for personal hygiene. Their content is quite amusing from the vantage point of today:

> Take a small bucket, and another container filled with cold water. You will also need a hot water bottle. Get hold of two sponges, like the ones sold for cleaning the floors, and one piece of flannel or wool. Place all of these on the table, and poor some of the hot water into the cold water to warm it up. Use the wool cloth to scrub the chest and the arms and anywhere you may have sweated. Then wash with both sponges energetically. (Vigarello 252)

The point of explaining the sponge bath was simply to show how even a poor family could afford to practice the basics of personal hygiene. On a continent where, since the Middle Ages, bathing had been rejected on religious and superstitious grounds (at least in the cities), there was a need to educate the whole population about simple rules of preventive health. The other venues, primarily schools, also served that purpose.

In the years of 1880–1900, the discovery of microbial infection led to a full adoption of the need for bathing. At the scientific level, the awareness of a need for clean hospitals and operating areas also became accepted. Once the causes of cholera, namely, a waterborne infection, became known, the outbreaks of such illnesses declined rapidly. Thus, medical discovery was of course central to that progress as was making the general population aware of the need for change.

WHAT'S IN A VACCINE?

French chemist Louis Pasteur came to developing human vaccines by first trying to cure animals. Having found a cure for silkworm disease based on microbial theory, he did not think of extending his findings to humans, even though a similar theory of disease propagation had existed in Parisian medical circles since 1861. He was trained as a chemist, not a medical doctor, and his path first took him through the study of animal diseases before he moved to investigate human illnesses.

Between 1877 and 1887, when he studied virulent microorganisms, vaccines did not exist. In ancient China, medical knowledge included the principle of limited infection for curing smallpox, but the process was often ignored because of its high risk to the patient. Pasteur initially accepted the contemporary notion that an infectious disease consumed elements in the body of the host. His work in biochemistry convinced him this was wrong.

Pasteur started working on understanding how illnesses were transmitted. He was not alone. In Germany, Robert Koch was working on isolating infectious factors, too, using animal tissue. Pasteur, using urine samples and boiling them, eventually found a series of three microorganisms, which in 1878 were named *microbes*, resulting from the contraction of *microorganism* and the suffix *bios* to stress that these organisms are alive. He then moved on to investigate the mechanism by which microbes spread and, following this, developed vaccines to treat animals, especially chicken cholera, anthrax (which affected sheep herds but also humans), and rabies observed in dogs.

Pasteur's interest stemmed initially from a wish to help the French government eradicate a serious economic problem: Once infected with anthrax, entire sheep farm populations disappeared, driving farmers out of business. After years of manipulation, which included the creation of a farm housing rabid animals for use in studies, Pasteur first confirmed that anthrax was caused by a parasite found in carcasses or as a spore in the soil (1877). The following year, his demonstration of microbial action helped establish the basic rules of asepsis or sterilization, first as a means to avoid septicemia. Other scientists, such as Joseph Lister in England, had already used this kind of knowledge to recommend transforming hospitals, in particular the surgical and obstetrics areas, into clean spaces.

Although Pasteur demonstrated publicly the value of vaccines on animals, he had yet to try it out on humans. Rabies, because it affected domestic pets and was deadly, was universally known. Few humans actually died of it, but lack of knowledge about its origins and workings mesmerized public imagination into viewing it as a paramount horror. Pasteur's scientific attention was first drawn to it in 1880 when a child who contracted the illness died. Pasteur began his investigation in December. He had his assistants focus on localizing the virus while he concentrated on exacerbating its virulence. In so doing, he came to understand the process by which the sickness started. It would take several years to devise an effective vaccine. The landmark case came on July 6, 1885, when a mother brought her son who was suffering from a bite by a rabid dog. Pasteur decided to undertake the treatment; he later defended his ethical choice by emphasizing how some fifty previous tests on dogs had proved successful. As the law forbade anyone who did not hold a medical degree from practicing medicine, the scientist called on a medical doctor to administer the injections. Three months later, the young boy was still alive, and new treatments began. Despite opposition from some in the medical establishment, by March 1886, some 350 people had received inoculations, and only 1 had died because the illness was too far advanced. Pasteur's greatest success now became the subject of reports worldwide, and the Pasteurian myth began to develop.

THE PASTEUR INSTITUTE

By 1881, Pasteur's reputation had spread around the world. Ordinarily, two attempts would be required before gaining admission to prestigious institutions, but Pasteur was elected on the first round to the French Academy. His reputation for infallibility gave him an aura that even collaborators were willing to consolidate: None ever criticized Pasteur's rush to implement the rabies vaccine on humans.

X-RAY RISING

Among the new discoveries with medical implications, Wilhelm Röntgen's study of what became known as X-rays had a phenomenal impact on the public mind. A professor of physics at the Maximilian University in Würzburg, Germany, since 1888, Röntgen became obsessed with a series of experiments that suggested what would later become known

An X-ray diagnostic table in 1896. The machinery was set up directly in the doctor's office and did not include any special protection for either the patient or the practitioner. Source: Courtesy Siemens-Pressbild.

as radiation. In 1895, though, he referred to *fluorescence* and eventually called these mysterious rays *X*, as they were akin to the unknown element in mathematics. The name took (though in German, the term *Röntgen rays* stuck). Once he was certain of his discovery, Röntgen shared the results by sending out X-ray plates. The most famous of these pictures represented Röntgen's wife's hand, showing bones and the wedding band. Röntgen won the first Nobel Prize in physics in 1901 for his discovery.

The potential for surgery was immediately clear, but setting up laboratories remained expensive and difficult. Few city hospitals prior to World War I could claim to have an X-ray photo laboratory available. When mobilization drew many scientists and technicians to the front, progress in spreading X-ray technology actually slowed down. In Paris, Marie Curie, the discoverer of radium and polonium, suggested the creation of mobile X-ray units to assist battlefield surgery.

Though proven useful, the development of associated fields of study (such as radioscopy and radiology, to train specialists) remained slow. Interestingly, about 150 women were trained in 1917 to act as X-ray specialists.

TWENTIETH-CENTURY CHALLENGES

Health conditions at the turn of the century showed limited progress. Whereas new strides had been made in medicine (notably through vaccines and the advent of clean areas in hospitals), on the streets sanitary measures that included the shifting of refuse waters from the middle of the road into an underground water system helped eradicate cholera outbreaks.

The medical world of the twentieth century is a study in paradoxes. While much of the progress and discoveries made in the nineteenth century continued to expand our knowledge of the human body and how to cure it, the ravages of war also helped advance other medical fields, notably plastic surgery, trauma, and psychological disorders. In addition, the impact on the patient involved new attention to individual needs. In parallel, though, the push for greater equality in treatment involved submission to new rules unimagined heretofore. Finally, medical discoveries, like scientific ones, spread much faster, and though their applications varied, one notices strong parallels between Europe and the United States.

Medically, the advent of the first clean operating rooms where sterile attire was to be worn and hands washed made its way into European society. By 1900, it was proper to avoid dirtying oneself in public. The advent of public urinals linked to and managed by the waterworks suggested hygiene in the street was also of concern, regardless of the presence of animal droppings on the streets. Yet, this was in line with the goal of reducing the number of infectious diseases that had affected Europe throughout the nineteenth century, especially the multiple cholera epidemics.

Government involvement in the health of its citizens was another characteristic of the scientific–medical world. France, for example, created its first health ministry in 1930. It enforced health regulations (such as the requirement of premarital medical exams). By the late 1930s, it had also made available maternal and well-baby clinics. The state supervision was thus likened to a paternalistic approach to individual health, whereby citizens complying with basic health requirements were offered further exams.

Prejudices about hospitals were also slow to disappear and remained strong until after World War II. Prior to that time, hospitals were deemed necessary as institutions caring for the poor. When the sophistication of medical procedures increased, so did the need to go into a hospital to complete the procedure (Prost 101). Women who might have had five children over a 15-year period may well have shifted from having their first four in the home to requesting a hospital birth for the last one. The reasons for this varied but included medical advice now given (and accepted) as well as the notion that this was the way one became modern (Saraceno 453).

THE FALL OF TUBERCULOSIS

The breakthroughs in medicine in the nineteenth century included a rethinking of how to deal with contagious illnesses. By the early 1900s, many countries responded to the medical discoveries by initiating new legislation regarding the spread of contagion. From 1900 onward, diseases were placed in the law books, and some were added to the list. The methods of control still included the age-old isolation of the sick but with moderation. Some illnesses, in particular venereal ones, were singled out separately, a reflection of social rejection of some forms of sexual contacts.

In Denmark, tuberculosis was included in legislation on infectious diseases as early as 1897, and in 1905 and 1912, special legislation allowing for compulsory isolation, compulsory treatment, and registration as well as free care was passed. Such extensive enactments were never passed for tuberculosis in Sweden. Legislation in 1939 included registration and compulsory testing but not compulsory treatment and isolation. Swedish politicians considered the use of compulsory measures potentially stigmatizing and believed that it would be counterproductive in limiting the spread of the disease (Vallgårda 108).

Until the introduction of the first vaccines in the 1920s, tuberculosis (also known as consumption because patients' lungs filled with fluid that forced expectoration) remained an important social disaster, striking all levels of the European population. By the turn of the century, paralleling developments in the United States, European nations undertook systematic efforts to reduce the spread of the disease. Over two decades, the rate of infection had fallen by 30 to 60 percent. Isolation measures were the primary weapon, followed by the use of disinfectants in areas known to have been exposed to tuberculosis patients (the saliva of which was contaminated). Another measure involved tracking the illness in cows, known to carry the disease: In France, for example, up to 50 percent of all cattle were contaminated in the interwar years. Pasteurization took care of the problem in urban centers, but the lack of similar measures in the countryside prompted the spread of numerous directives, such as boiling the milk for up to 20 minutes before giving it to children. Agricultural inspectors monitoring animal husbandry conditions in rural areas were also mandated to conduct tests in stables with the assistance of local doctors. Only once the tests had been confirmed as negative did the building receive official signage that meant the cattle, too, were safe.

In 1921, for the first time, a vaccine was discovered and successfully tested. Robert Koch in Germany had discovered the bacillus that caused tuberculosis but had been unable to devise an effective vaccine against it. Almost forty years later, Dr. Albert Calmette together with veterinarian

Camille Guérin isolated another bacillus, which they named after themselves and tested as a vaccine. The BCG, as it became known, was first tested on a newborn whose family members were exposed to the illness. Further tests, on monkeys and children showed positive results, with very few spreads of the disease. The BCG is a weakened form of the Koch bacillus and was eventually recommended for all children as a means to break the intergenerational transmission of the illness. Its adoption, however, was not systematic. Scandinavian nations and France required it, but other countries rejected its use, as it increased sensitivity to associated diseases. Nevertheless, the success rate in protecting children from deadly forms of tuberculosis reached over 90 percent. The concern recently has involved the rise of the illness in adult populations, as the BCG does not work well in adulthood. Tuberculosis happens especially in HIV cases where body resistance is reduced.

This did not stop all tuberculosis cases but reduced a majority of them. Adults and children already affected continued to make use of sanatoriums in mountain and sea areas, as it was believed that the air in such regions as well as exposure to the sun helped control the illness. It was the combination of public and private commitments, however, that helped control the spread of the disease. Yearly, European associations organized fund drives, either in the form of national collection or even of special stamp sales. Generally, though, by midcentury the most important forms of tuberculosis had disappeared from the European continent. This does not exclude that variations of the disease could reappear.

In the case of rabies, Pasteur's vaccine of 1885 was eventually replaced with a formula using inactive viruses. The disease in indigenous form has disappeared from most of Europe, though some cases are registered when travelers visiting areas where rabies still exists return home infected. Unfortunately, the vaccine is effective only in the initial stages of the infection. The obvious signs of illness for patients who do not know they are contaminated appear after the incubation period, usually too late: The patient eventually dies.

In the twentieth century, however, another application was found for the vaccine: inoculating animal populations. The virus remains active in fox populations, and in the late 1960s, reappeared in Western Europe, progressing some 25 miles per year. While attacks on humans were unknown, biting of cats, dogs, and cattle created a base for the disease to spread back to humans. The solution to the problem involved creating an attenuated live vaccine that was mixed in with meat and spread in areas where fox populations resided in greater numbers. The rate of success was about 75 percent, which meant that the virus was almost extinct (Hannoun 39). What this example shows is that the vaccination and prophylaxis process (that is the treatment of a contagion) does not end with the statistical disappearance of the disease but involves multiple steps to ensure it does not spread further.

In this respect, one would think that most European nations, many of which have instituted socialized medicine, would have a success rate equivalent to the eradication of many diseases. That is not the case, and it is the result of differing medical and sociopolitical approaches. In the case of measles, for example, Scandinavian nations adopted a systematic vaccination program akin to that in place in the United States, where a four-year campaign of information, inoculation, and prevention had essentially done away with the disease by 1982. Several central European nations, notably Czechoslovakia in the 1960s and 1970s, had achieved a comparable rate of success. Other nations such as France, England, and Scotland have not systematically enforced vaccination. Consequently, the rate has fallen as low as 70 percent for the inoculation of French children in the 1990s (Hannoun 113). Historically, only in cases where vaccination was legally required (and thus enforced for, say, entrance into primary school) have full success rates been achieved. The reasons for the limited success where no coercion exists involve costs (even where socialized medicine covers them) as well as misunderstanding of science.

THE INFLUENZA PANDEMIC

Though nowadays popular culture focuses on diseases that have a worldwide outreach thanks to modern transportation (such as SARS), it is important to consider that the first pandemic accompanied the horror of World War I and, in some regions, killed more soldiers than combat did.

The flu is a viral infection that was identified in the nineteenth century; various epidemics had swept countries then, but the Spanish flu (improperly named) became a pandemic. Current studies point to an avian flu developing in China around 1915 that reached North America through cargo ships and liners by 1918. Europe suffered the first cases that summer partly due to massive troop movements. Although soldiers were hit, civilian populations were the primary sufferers, for war conditions, especially food privations, had weakened immune systems.

Early cases were reported in Bordeaux, France, in April 1918, and the French government claimed Germany had introduced the illness though the infection of food cans imported from Spain. The disease spread in waves, first reaching northern France in May 1918. That same month, a separate outbreak was identified in Brest, on the Atlantic coast.

In every nation affected, health services were overwhelmed. In Geneva, Switzerland, for example, the peak of the epidemic happened in November 1918. Schools were closed, reopened, and closed anew while doctors, many infected themselves, faced overwork in trying to identify and treat the disease. They consequently cut their receiving hours, thus increasing the flow of patients in hospitals. Daily, the press published reports of new triage areas opened in requisitioned public buildings. Volunteers were asked to report there to help nurses, thus risking infection, too. As for death notices, they ran daily over several pages instead of the average half-a-page list (Ammon 190).

Under these circumstances, panic set in and rumors flew. Since all treatments seemed to fail, quacks went so far as to advocate alcohol-based remedies, eating onions, or getting plenty of fresh air exercise. Strange though it may seem, such reactions reflected the authorities' own incapacity in isolating the pandemic. Nobody knew how it was transmitted; some thought it was the result of dirty laundry washed by Red Cross volunteers, while others wondered whether letters soldiers sent to their families (the Swiss army was not fighting but was mobilized) from border areas had brought the illness home. The only concrete action to take was a statistical one, for the Swiss federal government ordered all cases to be reported at once and punished doctors who failed to do so (Ammon 191).

In some Swiss cantons, theaters, ballrooms, cafes, and even churches were closed. Funerary processions were limited to five people, and cemetery employees struggled to keep up with the tragic pace of funerals. Eventually, in Switzerland just as elsewhere in Europe, the disease subsided, leaving sorrow and questions in its wake. Not only had the strain of influenza yet to be fully identified but also how one could actually control the spread of another pandemic. The implication of isolation, for example, raised the matter of elementary civil rights in democratic societies. The question becomes a burning one in a case where the illness is not airborne, yet deadly.

THE AIDS EPIDEMIC

The American Centers for Disease Control (CDC) in Atlanta, Georgia, received the first reports of AIDS (acquired immunodeficiency syndrome) in 1981, but it was not until 1984 that the virus causing the illness was identified at the Institut Pasteur by the team of Professor Luc Montagnier (a later diplomatic agreement made this a Franco-American discovery). By then, what had first appeared to be an epidemic affecting primarily the gay community had spread to drug users but also hemophiliacs and anyone exposed to contaminated blood. Unlike other infectious diseases, however, the very nature of AIDS and the time of its appearance meant that it was both a pandemic with deadly outcomes and a mirror that reflected the tolerance levels, concerns, and ignorance of the societies dealing with it: AIDS is socially constructed, too, and involves as much stereotype and hype as it does factual and sobering information (Kirp and Bayer 1).

Awareness of the AIDS epidemic prompted all kinds of reactions across the European political spectrum and was also reflected at the public level. Stereotypical reactions akin to those recorded in the United States included dismissive claims that the homosexual communities alone were exposed to the virus. However, the fact that hemophiliacs as well as patients needing blood transfusions had become contaminated offered grounds for education rather than outright rejection. This shift was also the result of changing attitudes toward homosexuality, whereby acceptance or at least tolerance became more common in Western European cities. Such changes were slow, and this public ambivalence was reflected in national governmental reactions to the epidemic.

Generally, there are two main approaches to prevention of infectious diseases. One is a contain-and-control strategy, while the other involves a cooperation-and-inclusion approach. HIV/AIDS prevention efforts saw the use of both approaches (Vallgårda 100).

The first approach, which can call—and has—for mandatory testing, treatment, and even isolation, has existed for centuries and been applied to all infectious diseases. HIV/AIDS, classified as a contagious epidemic, thus joined in the 1980s other potentially deadly pandemics. The problem, however, was in the nature of the infection. The illness was not airborne, and, an infected person aware of his/her infection could help stop its spread. The other problem was that although HIV/AIDS had universal ramifications, its initial contamination realm was predominantly the gay community. To thus isolate and treat anyone known to be infected not only involved a breach of patient–doctor confidentiality (treated as paramount in Europe) but potentially would amount to preventively (and arbitrarily) isolating entire communities.

Parliamentary debates on the matter reflect the unease that surrounded an understanding of the disease. In the case of Denmark, the parliamentary law came to reflect a more liberal attitude toward the matter but also one that placed responsibility on the gay community to police itself through information. It also called for helping drug addicts with treatments. Indeed, aside from sexual contact, the use of contaminated syringes had been found to be the major culprit in spreading the sickness to the heterosexual population. The end result was that the authorities in Denmark chose an active role (information, treatment) but not one that might involve coercion (Vallgårda 105).

In Sweden, on the other hand, greater concern was placed on containing the epidemic and preventing its spread to the heterosexual population. While proportionally Sweden and Denmark experienced similar levels of HIV/AIDS infection (72 to 75% were infected by homosexual activity, while 3 to 4% got it from using infected needles), public discussion focused less on the gay community and more on preventing the spread of the disease through prostitution. The reasons for this are unclear, though it has been traced to a mix of greater acceptance of the gay community and the ability of the latter to take action once the problem became known. Drug use among prostitutes was widespread on the other hand, and the authorities felt that they could not reach the same level of understanding and cooperation as they had with gay associations (Vallgårda 109).

Other countries differed in their approach both as a matter of pragmatic concern for a general infection and also because of recent history.

The Federal Republic of Germany offers such a case in point. In 1983, the German government issued a statement designating AIDS a national concern and invited experts from the United States to brief German health officials on the matter. Though initially unconcerned by the spread of the illness beyond certain groups, by 1984 the government made AIDS a national priority. Population awareness was general within two years, and reactions fell into two categories, like those of officials and politicians: a minimalist approach (emphasizing information and voluntary control) and a maximalist one (calling for active control of populations at risk). Interestingly, media outlets did not follow an expected pattern of conservative press favoring maximalist control. In fact, populist magazines and newspapers chose human-interest stories, while some liberal-minded media either did not cover the tragedy or emphasized the need for control. As one analyst of this state of affairs suggests, this may have fed off and in turn excited public misunderstanding and hysteria regarding the illness (Kirp and Bayer 112).

Churches themselves adopted different stances. Protestant and Catholic denominations are the primary faiths in Germany. Protestant councils generally adopted an open-minded approach suggesting that one could not view the illness as punishment and had to help those affected. The Catholic Church on the other hand followed the papal notion that this was God's punishment for straying. Similar diverging attitudes prevailed in churches across Europe.

The illness had a legal impact, too, with courts being asked to decide whether a person infected with the disease who had knowingly had unprotected sex was in fact criminally liable (in the case of a GI stationed in Germany, he was). On the other hand, courts often came to favor the right to privacy and protection of the individual from any maximalist involvement on the part of the state.

This does not mean that governments remained uninvolved. Many countries, such as France, came to require the reporting of the disease. Although the nation in question was the first to have an epidemiological system in place for AIDS, it waited until 1987 to declare AIDS a "national cause" (Kirp and Bayer 226). This contrasted with AIDS awareness among the public and in the press. The death of famous personalities (such as philosopher Michel Foucault in 1984, but also American actor Rock Hudson, equally popular in Europe) of AIDS along with the open disagreements between France and the United States over who had discovered the AIDS virus first became front-page news. Still, it was not until 1988 that the French public was introduced to the first condom campaign that argued AIDS was not a socially restricted epidemic. By 1991, the French government intervened to ensure that HIV-infected patients would not be asked about their private lives when seeking insurance but would be considered higher risks for insurers, like sufferers of cancer were (Kirp and Bayer 245). Overall then,

France responded slowly to the disease but eventually adopted a principle of social solidarity that relied on consensus of expertise to be effective.

The AIDS epidemic also brought forward a new wave of discussion about human sexuality. It was only the latest in a series that dated back to the nineteenth century and involved contraception.

SEX AND BIRTH CONTROL

By the start of the modern era, infanticide was a common and tragic outcome of unwanted pregnancies. However, numbers are hard to estimate, as accidental deaths also occurred. Historians estimate that the use of drugs, ingestion of alcohol, or simply too little food often ended a baby's life. Some parents, based on constable records, were known to have accidentally smothered their children in bed. This is why, combined with the new awareness of children as special beings (as argued by philosopher Jean-Jacques Rousseau and others), several foundling hospitals were established throughout Europe to help alleviate the trauma of infant death or abandonment (Potts and Campbell 27). Many of these in fact received children of legitimate birth.

The nineteenth century witnessed several innovations in birth control, but social mores limited the effectiveness of new practices. These varied considerably from one region to another and according to confessional preferences, too.

Strikingly, illegal abortion became a standard way of dealing with the problem of unwanted pregnancy. Back-alley abortions were offered through newspaper announcements that appealed to women suffering from "a delayed period" or who were simply temporarily indisposed. The methods used involved homemade medications intended to provoke violent illness in the patient, which would result in spontaneous abortion. Sometimes, accidental knowledge helped the provider. Witnessing that many pregnant women had aborted in Sheffield in the late nineteenth century due to an outbreak of poisoning linked to lead pipes in the water supply, providers set about handing out diachylon. The planters contained lead and acted as an abortifacient. In 1898, the first such case was reported in the same city, but it took almost 20 years before British authorities placed the plaster on the list of poisonous substances.

In the French Revolution, for example, the manufacture and sale of condoms, which had existed for centuries (made out of animal guts), was legalized. Of limited effectiveness, it was nonetheless deemed useful for men who wished to avoid fathering a child out of wedlock. Improvements such as latex condoms did not appear until the 1870s, almost four decades after Charles Goodyear had discovered and patented latex. The cost of such contraception was such that after each use it was to be washed, dried, and covered in talcum to be reused.

Though Europe experienced a population explosion lasting up to the mid-nineteenth century, by 1900, several nations were in demographic decline. No single explanation has ever proved satisfactory, and historians suggest that only a combination of circumstances can help explain such a shift. In the case of Great Britain, the decline extended almost 70 years, from 1870 to World War II. The first factor may have been better education. Though Victorian England abhorred any notion of "filthy" discussion, several pioneers of sex education made their mark there. John Stuart Mill (1806–1873), as a free thinker, advocated birth control and was even arrested for handing out leaflets on the subject. Englishman Robert Dale Owen wrote *Moral Physiology—or a Brief and Plain Treatise on the Population*, which was published both in the United States, where he had immigrated, and in England. Soon after, Charles Knowlton published *The Fruits of Philosophy* anonymously, which discussed various methods of contraception. In the 1870s, the British court case *Rex v. Charles Bradlaugh and Annie Besant* charged two defendants with republishing the book (which had been censored). While the jury found the book immoral, it could not find corrupt motives against the couple, which had freely advised the police of their actions to bring about a discussion of the need for birth control.

What is interesting in all these cases is the reaction of the medical profession: It opposed all forms of birth control, and it echoed such views through the prestigious publication *The Lancet*. Echoing the medical profession, the churches strongly condemned any form of what they viewed as the road to recreational sexual relations.

Perhaps the most important figure in the saga of birth control in England is Marie Stopes (1880–1958), as she made any who would listen to her aware of mechanical birth control methods. Stopes, an ardent feminist, published her second book on the theme of birth control after having met Margaret Sanger, a birth control advocate from the United States. The danger was substantial, as several of Stopes's predecessors had ended up jailed for advocating such a position. Stopes's book *Wise Parenthood* shocked church leaders, but conservative groups attacked her using obscenity laws. Though she was able to fend many off, it was clear that she faced an uphill battle in making readers aware of how basic birth control methods offered social benefits in the longer term. But Stopes was a small factor in a sea of change. Suggestions that new attitudes toward sex came after World War I when women ended up working in factories are taken at face value yet do not account fully for the continuing decline in births even after soldiers came home. Others yet think women simply took better control of their sex and family lives as marriage moved away from arranged patterns toward more consensual ones. Regardless of which factor was the most important, recent research suggests women were generally ignorant of such factors (Fisher 70–72), partly because any discussion of sex was either deemed an affront to private life or outright obscene.

This state of affairs prevailed until after World War II. One reason for the opposition to birth control, besides religion and social tradition, was political. In France, for example, a natality policy dating back to the population decline of the early twentieth century had become a cornerstone of national polity in the interwar years. It thus comes as no surprise that when the pill first appeared in the United States in 1960, whereas some European countries authorized its commercialization, France remained steadfastly opposed to it, prompting a few French women in Grenoble to create a clandestine family planning center in 1961: They would later explain how pill boxes smuggled in from England and Switzerland would be dispensed as a means to regulate the period, without any mention of the contraceptive properties. Finally, in 1967, the Neuwirth law cleared the way for the commercialization of the pill and its dispensing in government planning centers. The opposition of many lawmakers to this decision delayed the actual publication of the law until 1974, the same year abortion was legalized there. Abortion legislation, when it allows the procedure, diverges widely from one nation to another. Many restrict the practice to first-trimester pregnancies and require a 24-hour waiting period following a counseling session (Switzerland). Other nations require parental consent. One element common to most is the availability of information on birth control and on sexually transmitted diseases, either in public health centers, or through schools.

CONCLUSION

We have seen how the gap between scientific knowledge and public understanding of the body, diseased or healthy, actually affected the way illnesses were handled in modern European history. In particular, contagion when associated with sexual mores has influenced heavily the process of education, information, and control of illnesses. This is unlikely to change as illnesses themselves become far more globalized than ever before.

II

SCIENTIFIC
CHALLENGES

The modern era in the sciences witnessed a dual phenomenon that on the one hand popularized science in the nineteenth century, thereby making it intelligible to educated enthusiasts and spreading scientific knowledge through the media. Then also, the process of formally educating future technicians and scientists also took place. In the twentieth century, however, a new shift changed the way science operated. First, the globalization of scientific endeavor, begun in the preceding century, became far greater and complex. Scientists trained in Europe might in fact not make their greatest contributions until they began work elsewhere, often in the United States. In addition, the impact of the world wars caused a pessimism in the scientific and technical realm that had not existed previously. Finally, whereas science had been popularized in the nineteenth century, in the twentieth, its applications, though enjoyed by millions, became intelligible as the bodies of knowledge grew in complexity and depth.

Although the most important scientific revolution in 250 years had its roots in the nineteenth century and affected few in everyday life, it is worth summarizing, partly because of the impact it had on the popular cultural perception of the scientist.

In the seventeenth century, Isaac Newton had provided the capstone to decades of new discovery in astronomy, mathematics, and physics by positing a series of laws to explain the functioning of the universe. In so doing, he removed religious belief from the equation, and his and others' subsequent discoveries played an important role in both the Enlightenment and in the backlash against the church. Newton's laws were clear, and any discrepancies in the observation of the stars, for example, were overlooked. Several discoveries would change that.

The first change in the physical outlook on the world involved the discovery that atoms were not the smallest elements in the universe, a theory posited as far back as Ancient Greece, but that particles named electrons formed them. Further research followed at both the theoretical and laboratory levels in the United Kingdom and continental Europe, whereby the phenomenon of radioactivity was posited. In addition, two new theories appeared.

The first was quantum theory, presented by Max Planck in December 1900. In it, he suggested that energy is not emitted in various quantities but in standard packets. The fact that energy might not flow constantly contradicted Newton's notion of a steady flow of energy.

The second theory concerned relativity, and like quantum theory, the extent of the revolution it caused is still under investigation. Instead of space and time being constants, light speed, based on Einstein's work as well as the Michelson-Morley experiment, was a constant. Einstein went further by positing first that gravity could affect light, in effect bending it. By suggesting this effect (which was proved through a British observation of planet Mercury in 1919), Einstein changed celestial mechanics as devised by Newton, putting away for good the notion that ether was a material that inhabited space. Most importantly, Einstein proposed his now famous formula $E = mc^2$, which suggested the possibility of converting matter into energy, with important energy expansion if material were radioactive and fissile.

Other theories and discoveries followed. Their relevance in everyday life is that, for a time, science became the stuff of daily media coverage. In parallel, however, the general public understood less and less the actual scientific discovery and its implications. For example, Einstein's theory of relativity was deemed "understandable to only a few" according to reports in the popular press, while intellectuals grew concerned about some of the corollaries of a discovery of a cosmos that was seemingly unstable and definitely not in tune with an established order. This sense of uncertainty contributed to some scientists seeking solace in the support of authoritarian movements: The chaos caused by World War I and its aftermath as well as the seeming acceleration of change in society made the call for enforced stability appealing to some who longed for an illusory pre–World War I past.

Several of these processes took place along parallel tracks, and to survey all would require an encyclopedic approach. Therefore, this chapter offers examples intended to illustrate how scientific knowledge evolved over two centuries by emphasizing the notion that contrary to the common belief of constant progress, the exact sciences, like other forms of knowledge, also backtrack and face dead ends or outright opposition to the results they produce.

MAKING THE SCIENCES INSTITUTIONAL:
A NAPOLEONIC EXAMPLE

The loose conglomerate of scientific thought in the Enlightenment (see preface) meant that each country would have unique experiences that would shape the formation of institutions of higher learning. Most stories of how these institutions came to be involve a solid degree of determinism and positive outlook. The rise of nationalist feeling in the nineteenth century played a role, of course, but so did the need to convince the public of the value of scientific study.

Reality shows that there is no straight path, and the example of how France laid the foundations for some of its best schools makes that clear. Ideas of the Enlightenment eventually found expression in the break the French Revolution brought about. To carry out to fruition the goal of reforming French culture and institutions, revolutionaries sought to spread knowledge far and wide. The Louvres Palace, for example, became a national art museum open to the public. New scientific institutions also appeared. What the revolutionaries began, Napoleon Bonaparte completed. Controversial for his military career, several social and cultural changes he instituted would have a lasting impact on institutions of higher learning and the eventual development of scientific programs in universities

SCIENCE UNDER NAPOLEON

Napoleonic France's contribution to science consisted of crystallizing changes begun in the French Revolution and building upon these. In so doing, the emperor carried on the tradition of centralization that had already existed during the *ancien régime*. The concentration of scientific talent in Paris was partly the result of revolutionary upheaval, when many scientists had joined the *Levée en masse* decreed in 1793 to defend France against pro-royalist armies. Consequently, the huge number of scientists present in a relatively small area encouraged exchanges of ideas and creativity.

As a young man, Bonaparte developed a general interest in science and a facility for mathematics, which pushed him into artillery training during his time at the *Ecole militaire*. From then on, he made the acquaintance of and studied (either formally or casually) with mathematicians, physicists, and natural scientists. Several became his friends, including Gaspard Monge (the father of descriptive geometry who helped found the *Institut national* in 1795, later the Institute of France), whose brother had trained Bonaparte in mathematics; Laplace, who was his school examiner; and Berthollet, whom the young general had met during the Italian campaign in 1796. All three supported Napoleon's admission to the Institute. Although the officer's publications suited him best for a post in the Second Class (moral and political sciences) or Third Class (literature and fine art), he desperately wanted to be a member of the First Class (sciences). Despite tough competition, he was successfully elected, a move described as a victory both for him and for the Institute, hoping to incur political favor with the revolutionary regime.

Although Bonaparte attended several sessions of the Institute, his military duties took him away from France soon after his election. He did not forget his interest in science, however. Starting in 1798, when the notion of an Egyptian campaign was under discussion within the Directory, Bonaparte began recruiting the best French scientists of the day and even succeeded in convincing middle-aged men such as Gaspard Monge and Nicolas-Jacques Conté to accompany him to Egypt. Although he named himself a member of one of the scientific committees in Egypt, Bonaparte did not, however, make any presentations at scientific gatherings, although he joked with some of his scientist friends that he ought to do something soon. His role, however, as a morale booster was just as important. Landing in Egypt at the same time as the French army, the scientists encountered heavy hardship in adjusting to weather, hygienic, and building conditions. Months passed until a pleasant villa could be secured to house the Institute and its members, during which Napoleon regularly met with scientific leaders and promised to make the necessary logistics available.

Monge, Berthollet, and Magallon acted as temporary administrators and oversaw the structuring of the Institute, with regular meetings scheduled to allow members to report on their projects. The scientists were paid according to a specific pay scale set up in fall 1798, ranging from 500 livres for "savant de première classe" to 50 livres for a "savant de 10 ème classe." The Institute put engineers and geographers to work on projects of immediate use, such as drawing a 1/30,000 scale map of Egypt under the leadership of Cafarelli. Meanwhile, Nicolas-Jacques Conté, among his many accomplishments on Egyptian soil, set up a factory to manufacture material for French army uniforms and later presented a time-measuring machine at the November 12, 1799, meeting. Gaspard Monge for his part, investigated ancient world history by tracing the Roman canal that linked the Red Sea to the Nile River. Others focused more closely on Egyptian civilization, studying, for example, the recently discovered Rosetta stone. After Bonaparte's departure, the *Institut d'Egypte* continued to function until the arrival of the British.

Homesick, Monge was most pleased to return to France with Napoleon and assume the position of director of *Ecole polytechnique* (established in 1794 as a means to help French industry with a strong scientific foundation). He continued to support Bonaparte actively and advise him on political and scientific matters until the latter's fall.

When Napoleon took control of France, the organization of scientific knowledge was divided into three parts: scientific research, science teaching, and training in engineering and medicine. Several scientific research centers thus thrived, like the *Collège de France* and the *Muséum d'histoire naturelle* (previously known as the *jardin du Roi*). Napoleon's contribution consisted in restructuring scientific research and teaching.

Napoleon decided to extend the teaching of science beyond specialized schools and into the university. The Napoleonic system favored six major disciplines: botany, chemistry, geology, mathematics, physics, and zoology. As first consul, he emphasized the importance of furthering the sciences by offering several monetary prizes. The first was an award to the Italian scientist Volta on the occasion of the latter's visit to Paris to lecture on electricity at the Institute, which was followed by a yearly prize for work on electricity. Several other prizes followed, reflecting Napoleon's agenda of favoring cutting-edge work in the sciences and technology. While such practices were not unique to France, their number and compensatory amount were.

This does not mean that all scientists and inventors incurred success. American Robert Fulton tried to attract French interest in his *Nautilus* submarine (which he tested on the Seine River in 1800) and later in his steam boat (1803). While he did get a government grant, the lack of efficient propulsion prompted withdrawal of French support, and Fulton returned to the United States.

Research was entrusted to newly centralized universities, the *Ecole polytechnique* and the *Ecole normale supérieure*. The *Ecole polythechnique* offered systematic training in engineering and the opportunity to learn in a research laboratory. Major French scientific figures such as Monge and Laplace thus taught there while pursuing research. Although it combined the fields of science and engineering, it evolved slowly toward teaching more of the latter as a means to prepare men for military duty.

Napoleon's patronage of the sciences was often subordinated to his military goals. Under the Consulate, the *Ecole polytechnique* served as a surrogate military academy, and in 1804, it was completely reorganized along military lines (despite protests from Monge and others). By 1811, future officers were ordered away from it straight into military academies, while students not planning on a military career, yet who finished at the top of their class, would be drafted as military engineers.

The *Ecole normale supérieure*, derived from a short-lived 1795 experiment, was organized in 1808 to supply the teachers required for the new national centralized secondary education system. This system culminated in a series of examinations, the *baccalauréat*, a standardized exam designed to ensure that a uniform education had been imparted to all. That same year, the *Université de France* (or Imperial University) came into being, bringing French universities, until then independent, under control of the state.

Napoleon's establishment of the Imperial University system was an attempt to stifle conservative opposition by consolidating civic unity. This institution, set up by 1808 and 1810 decrees, was to oversee an educational structure that began with the lycée and included the municipal college (communaux) and the different faculties. Some 15 faculties of sciences came to exist in Napoleonic times. Intended to diffuse knowledge, the model adopted raised contemporary criticism from Saint Simon for instilling a classics-based culture that stifled innovation by simply certifying students for particular jobs. In particular, the system called for specific degrees to hold specific appointments. The doctorate was essential to a university post, while teaching in a lycée could actually require an Aggregation, a rigorously competitive exam, which, although sporadic in the eighteenth century, now became a standard requirement.

In the field of medicine, the French Revolution had dissolved the Royal Society of Medicine, and no regulatory system appeared in its place because of the notion of freedom of profession and associated "medical liberty," thus leading to an overabundance of doctors trained in diverse and unsystematic manners. The urgent need for medical doctors to help the revolutionary armies reinstituted formal training in 1795, but it was not until 1803 through the *Loi de Ventôse* that the state stepped in to license individuals to practice medicine, thereby displacing the old corporatist model that had dominated the profession until the French Revolution. The image and function of medicine thus began to change, but the primary function of medicine in warfare likely delayed interest in formally reorganizing the profession.

Napoleon's impact on science was also the result of a movement that began in the Enlightenment and flourished in the French Revolution. Despite strict political control, a strong intellectual life continued to thrive under Napoleon and confirmed the utilitarian vision the emperor had for the sciences. The Institute, for example, continued to be recognized as a place where scientists might advise political leadership, and they were paid as public servants. The Institute was also to spread the results of individual research. However, Napoleon did interfere in its functions by forcing a reorganization of its structure. As of 1803, rather than three classes, four were introduced. While sciences remained the First Class, French literature made the Second; history and ancient languages, the Third; and fine arts, the Fourth. Moral and political sciences "disappeared" because, in Napoleon's view, they constituted a potential challenge to his power. His favoring of the sciences was also clear in the amount of the yearly prize awarded at the Institute: 3,000 francs to the best project in the First Class, versus 500 for the Second.

Scientific life under Napoleon was predominantly centered around Paris. Other institutes included the *Collège de France* and the Natural History Museum. Other institutions existed that allowed for a more flexible interaction among scientists, such as the *Société d'histoire naturelle de Paris* or the *Société d'encouragement pour l'industrie nationale*. In addition, a smaller group, the *Société d'Arcueil*, was first organized by Berthollet and Laplace and gained in importance as of 1807, when it began publishing a journal. A loosely organized association, its informal atmosphere provided a pleasant alternative to the stricter testing grounds of the First

Class of the Institute, of which several founders of the *Arcueil* group were members.

Many French scientists distinguished themselves under Napoleon despite the latter's focus on the practical aspects of science. For example, Berthollet published an *Essai de statique chimique* (1803), which applied Newton's theory of chemical affinity. Gay-Lussac discovered the laws of expanding volumes of gases, while at the Muséum of Natural History, Lamarck theorized about a transformationist theory that Darwin would study later to devise his own theory of evolution.

The main impact of Napoleon Bonaparte's involvement in the sciences was to centralize them. Such state control made for remarkable progress within state-sponsored scientific circles, but it also meant a restriction of funding outside of such spheres. The patronage of Napoleon also confirmed French domination in the sciences as had begun in the mid-eighteenth century. However, other shortcomings soon were added to overcentralization, including the fact that there were too few outlets for the number of newly trained scientists that the Napoleon-instituted system started churning out. Nonetheless, several of Napoleon's actions laid the foundation for further developments in European science and education and for France's industrial transformation later in the nineteenth century.

EDUCATING THE ENGINEER AND SCIENTIST

The nineteenth century witnessed the professionalization of several scientific and technical fields in a manner that is still reflected in European educational tracks nowadays. Early twentieth-century science benefited tremendously from the emphasis national governments had placed in previous decades on the need for building an educated workforce. Applied engineering, but also specialized schools in physics, biology, chemistry, and medicine, became the norm. Universities that managed these centers also competed for the best minds.

SCIENCE AND RELIGION

Among the biggest transformation of science in everyday life in Europe was its reinterpretation into a challenge to organized religion. The departure here is worth noting because, until the eighteenth century, most noted scientists practiced their faith. Isaac Newton, but also Robert Boyle (an English pioneer of chemistry), openly stated that their work was merely an attempt to understand the great work of God. The shift began in the Enlightenment, as a reaction against churches' political abuses and as historical examinations of clerics' actions against scientists recast the importance of theological knowledge in relation to scientific knowledge. It is in nineteenth-century Europe that the debate gains in importance. Social Scientists like Karl Marx, but more importantly developers of new

Engineers in the lab, 1905. This picture of engineers at work in the Berlin branch of industrial concern Siemens & Halske reflects the substantial advances in applied science at the turn of the century in Europe. Source: Courtesy Siemens Pressebild.

scientific fields, such as Freud and Darwin, prompted a questioning that went beyond simple rebellion. Whereas the preceding wave of challenges to churches had involved a focus on the human—therefore fallible—side of religion, this time the conclusions many drew from new findings pointed to God as a simple construct.

This did not occur in all scientific fields. Augustinian monk Gregor Mendel, living in Austria Hungary in the 1860s, described the foundation of heredity all the while officiating in a monastery. Darwin's new theory, however, challenged far more than heredity.

In 1858, a scientific commission of the Linnaean Society of London learned of the work a young naturalist named Alfred Russel Wallace had compiled. It echoed earlier endeavors another scientist, Charles Darwin, had undertaken and almost displaced his own work. In fact, the former agreed that the latter had progressed beyond what was contained in Wallace's work. Yet, Darwin's own conclusions did not seem to be firsts, either. At the turn of the nineteenth century, several scientists, notably Jean-Baptiste Lamarck had posited the notion of evolution based on primitive organisms. Few, however, shared this view, and many continued to follow the biblical premise that God-created species never changed. The only way they might disappear would be due to environmental changes. When Darwin had traveled on the HMS *Beagle*, he had followed part of the latter premise, but after his return, he began to reexamine his own understanding in light of what he had observed on his five-year trip. He then rushed *On the Origin of Species* into print in 1859. Where he departed from Lamarck and others was in the evidence he offered as to *how* things evolved. By emphasizing randomness in the process of survival and evolution, Darwin's conclusions implied that humans might not even be the highest evolutionary form. Variations tied to the environment as much as to the species involved resulted in the struggle for life, which so many identified as the crux of the Darwinian argument. If one applied this to humans (Darwin would not do so in his first volume), then the soul itself might be viewed as the product of an evolutionary pattern. Despite the meticulous wording and structure Darwin applied to his work, the stage was thus set for a series of public debates.

SOCIAL DARWINISM AND THE RISE OF SCIENTIFIC RACISM

Contrary to popular notions of the reception of *On the Origin of Species*, there was no "Darwin Affair," as such. Some very public and forceful debates took place, but British society in particular eventually followed Darwin's argumentation. A substantial debate at Oxford University in 1860 between Thomas Huxley, a defender of Darwin, and Samuel Wilberforce, the bishop of Oxford, ended up in the majority of the 1,000 spectators supporting Darwin. The storm in the teacup did spill over. On the European continent, there was little reaction against Darwin's ideas, too. His publication in 1871 of *The Descent of Man* did, however, start a different kind of controversy, which Darwin had not expected.

In 1877, the Catholic Church reacted against Darwin's work. Pope Pius IX issued an encyclical that evoked "the aberrations of Darwinism." What is interesting in the papal message is that Darwin himself is not condemned but rather specific elements tied to the understanding of his theories.

Englishman Herbert Spencer (1820–1903) and many of his followers misunderstood Darwin's theories and posited that not only was there a notion of progress tied to evolution but that humans were affected by an inequality in their traits and characters. Philosophically, this meant that whether one could improve might be less tied to personal drive and education than to physiological traits. The implications of this interpretation (which Darwin rejected) continued to be discussed at the political level, with chilling implications: The notion of survival of the fittest that Spencer and others posited fed racial science, a pseudo-scientific field that used scientific rhetoric to argue generally for the superiority of white Europeans over other races. In so doing, it also introduced a new form to an old hatred. Racial anti-Semitism borrowed liberally from Social Darwinism to argue for the need for racial cleansing. Much of this was expressed in the eugenics movement, which spread widely in Europe and the United States.

The modern idea of eugenics originated in England with Sir Francis Galeton, who helped found a British society to study eugenics. A sister organization to the British group, the American Eugenics Society was formed in the United States in 1935. The implicit belief of eugenicists was that races were genetically superior or inferior and that to mix races meant putting "pure racial stocks" at risk. Since then, scientific evidence gathered through the genetic study of plant observation suggests the contrary result to be the case.

In the early 1930s, the concept of sterilizing some members of the human population was commonly accepted in 27 states of the American Union, although several eventually withdrew the legislation authorizing the practice. The idea behind the practice was that to succeed in building a strong nation, social engineering, or so it was believed, had to extend into controlling the human reproductive cycle. *Scientific American* did not hesitate to proclaim that "one-fifth of the population of the United States today is surplus" (Landman 295), noting however that the eugenics movement had not yet proved its case for full population control. Such emphasis, however, did little to temper the partisans of full-scale eugenics.

Up to 1933, the American eugenics movement displayed strong power in influencing domestic legislation concerning race and racial hygiene, and American eugenicists received praise from Europeans, especially German eugenics advocates. By the time the Nazis came to power, it was believed that the two leading eugenics movements were in the United States and Germany. Nazi Germany purported to follow the precedents set by American sterilization policies, especially the California sterilization law. Soon, however, it is Nazi propaganda that took the lead in explaining the benefits of sterilization for a "purer race." Furthermore, Nazi policies extended to the whole German nation, while eugenic laws in the United states encountered road blocks at federal, state, and even local levels. The Secretary of the American Eugenics Society, Leon F. Whitney, reported on a regular basis on the progress of Nazi policy. It became clear to American eugenicists that the German treatment of the Jews was no different than the American treatment of blacks and was, according to them, acceptable. However, as German measures against the Jews radicalized, relations between the American and German eugenics movements cooled down considerably. In fact, Nazi abuses of eugenics in the name of anti-Semitic policy tainted the very term and may have contributed to the toning down of American rhetoric in the field. The rise of genetics as an established field of biology also dispelled eugenic myths.

NAZISM AND SCIENCE

When Hitler came to power in January 1933, he inherited the world's most advanced scientific community. German technical and scientific schools had shone since the nineteenth century, with some of the major discoveries in chemistry (Wilhelm Ostwald and Fritz Haber), physics (Planck, Einstein, Born, Schrödinger, Heisenberg), and medicine (Koch, Roentgen). This also applied to engineering (Bessemer process, Daimler and Benz). What had allowed the research and university system to thrive was its relative independence. Despite the fact that scientists and engineers working at universities were technically paid bureaucrats, they enjoyed great autonomy of research, thus enabling them to get involved in long-term projects that might not yield results for years. This all changed as Hitler consolidated power. He first had passed the Enabling Act in April 1933, which provided for the dismissal of any non-Aryan scientist or any potential opponent to the system. Einstein had already resigned, but others still in Germany found themselves threatened.

Fritz Haber, a nationalist scientist, was a protestant of Jewish descent and thus suspect. However, because he was a World War I veteran, he was exempt of the resignation requirement (Hindenburg demanded this for his veterans). Yet, Haber resigned on April 30, 1933, because he had lost so many staff members, he felt unable to work effectively. The end of his letter of resignation is quite telling in this respect: "In a scientific capacity," he wrote, "my tradition requires me to take into account only the professional and personal qualifications of applicants when I choose my collaborators—without concerning myself with their racial condition" (Hentschel 44).

Dr. Haber's administrative superior was Max Planck, a physicist who devised the early development of quantum theory. Planck's reputation in Germany was absolutely stellar, and he tried to use this respect to intervene on behalf of Haber. When he approached Hitler, the latter went berserk, screaming: "Our national policies will not be revoked or modified, even for scientists. If the dismissal of Jewish scientists means the annihilation of contemporary German science, then we shall do without science for a few years" (Hentschel 360).

The consequences of such an attitude soon became clear: Germany would loose a total of 7 Nobel prize winners in the sciences and another 13 top scientists who would win Nobel prizes in years to come (total: 20 Nobel prizes). Hitler never felt any remorse. In fact, he went about attacking the objectivity of science, arguing it was a slogan and that, in fact, science had to be guided by specific ideals. Thus, to Hitler, there was a Nordic science and a National Socialist science, working together hand in hand. These fought what he called a "liberal Jewish science," which was bound to be destroyed.

Symbolically, the book-burning session of May 10, 1933, reflected this decision to remove certain scientific approaches: Many of Einstein's books, for example, went up in flames on that day.

With that in mind, it becomes necessary to understand the ideology of national socialism. It was derived from the post–World War I Fascist movement, which stresses the importance of the state over the individual and the rejection of civil rights. It is against the church and against democracy. It functions based on (1) strict hierarchy, (2) authority/decision by command, and (3) discipline that subordinates the individual completely.

Nazism added the process of *Gleichschaltung* and *Voksgemeinschaft* (coordination and community), which means in effect the implementation of cells all subordinate to the party, be it a teachers' union, the beekeepers' society, or a scientific association.

To continue to function in German society, non-Jewish Germans were expected to conform but not necessarily to become party members. Many chose to join nonetheless, especially in 1937, when the membership rosters, sealed in 1933, were reopened.

The results of such control over science had, of course, a horrible impact on not only the medical profession but also the physical and biological sciences. Among the biologists who remained in Germany, 53.2 percent of the sample joined the Nazi Party, a number lower than in the medical field but comparable to that of psychology. The reasons for joining appear similar to those in other learned professions—professional advancement. Yet, if membership in a Nazi professional organization was the norm, and often required to remain in a given profession, party membership itself was not.

Sadly, several Nazi biologists, including Lehmann, were reinstituted in the 1950s despite strong protests from their colleagues. A most glaring case is that of 1973 Nobel Prize winner Konrad Lorenz, whose ethological research on animal behavior interested the Nazis for its potential application to humans. Although Lorenz's work established a bona fide biological field, his comments during the period 1941–1945 suggest that he should have been excluded from further activity.

During the war, scientists often camouflaged basic research under the guise of work essential to the war effort. This ensured that they would obtain an SS classification for funds and materiel appropriation. Yet, many such projects had little, if anything, to do with the war effort. The habitual contradictions between National Socialist ideology and the war machine are also visible in the projects that received support. The *Deutsche Forschungs-Gemeinschaft* (German Research Association/DFG), for example, refused to fund studies of prehistory until it found out that Hitler himself had charged the leader of the Prehistoric League with the task of working on these issues.

What these cases showed was that besides the obvious horrors associated with the misuse of science, the reaction to these was often muted, as if science were an apolitical tool. The reality, as shown here, was clearly otherwise.

THE POSTWAR ERA

Quantifying the postwar era is difficult in the context of a brief survey. Generally, Europe took a second seat to the leadership the United States established in World War II. Budgets as well as training contributed to a lag in the development of scientific discovery, though applied science continued to fare well, as in the case of nuclear power. There, pre–World War II work on the subject as well as U.S. assistance allowed the slow, steady establishment of a nuclear power tradition.

Europe eventually regained a position of strength in chemical, pharmaceutical, and engineering realms. Consequently, scientific research did go through a period of growth.

One should also note that the traditional educational system in several countries actually did not recognize the study of psychology and sociology as valid fields until students passed graduate exams. This changed in the 1970s, but it reflects how Europe had fallen behind in several scientific fields. Other factors in the slowdown included the difficulty students encountered in transferring credits from one university to another, even within a single country. Undergraduate education thus extended easily to seven years on average for a three-year program. Though the problem did not exist on the same scale at the graduate level, the limited number of spaces available in specialized laboratories also meant that some research experienced slowdown.

Another factor for the European slowdown was the matter of funding. Whereas in the United States industrial investment allowed for substantial support of doctoral and postdoctoral candidates, in Europe, limited national funding to universities and industry had a substantial impact. Contemporaries spoke of a so-called brain drain that shifted the best minds to the United States starting in the late 1960s and drawing the best scientists until the 1990s. Other factors involved marketing failures in the case of engineering innovations and limited economies of scale for the commercialization of new products. Some institutions naturally were able to maintain national leadership in their respective nations, but many did so by emulating models of graduate education found in the United States. This in turn may have facilitated a notable upturn in international cooperation. Exchanges through such organizations as the Fulbright scholarship program opened new channels of knowledge exchange, which helped Europeans maintain efficient contact with American institutions.

CONCLUSION

The cycle of scientific evolution during Europe's modern period is staggering. From a small field of study for a few free thinkers and avid enthusiasts, science underwent a formalized approach to training, all the while appealing to a wider educated public. Inventors in particular became new heroes offered up in history books on a scale comparable to great conquerors (and often using the same rhetoric for struggle and success). Yet, the strides science made also created a chasm between the scientist and the public: Fewer understood what scientists actually did.

The misunderstanding was eventually exacerbated through the horrors of the world wars and, in many ways, found its ultimate expression of mistrust in the environmental movement that started in the United States yet spread widely to Europe in the 1970s and 1980s. By the late twentieth century, however, a new swing in the balance was occurring that took into account the need to educate the general public and rebuild bridges to the scientific profession. Though not new, the rebirth of a public interest in science is part of the globalization process that has witnessed science becoming an intercontinental event rather than a national or even European endeavor alone.

12

CONCLUSION: A GLOBAL EXPERIENCE?

There are hundreds, if not thousands more technical contraptions and scientific methods that have impacted the lives of Europeans since their continent entered the modern era. In the context of everyday life, some gained in importance because of their function, their value, or even simple marketing and word of mouth.

The standard-issue Swiss army knife, produced by Wengen and Victorinox, comes to mind. The corkscrew was added to satisfy the civilian market. Any other Swiss army items, however, have not seen any kind of military service and are more likely to be sold outside Switzerland. The point is that by the early twenty-first century, the globalization process that had been ongoing for centuries now accelerated and thus blurred boundaries of what is inherently used in Europe or in a European way. In fact, the Swiss army knife is no longer a tool of function for men in uniform cleaning their rifles or shoes. It is part of the leisure equipment of travelers.

Technology indeed became as much an instrument of leisure as of necessity. Some scientific, but especially medical, developments also have brought about the issue of choice in case rather than necessity, and the social and medical structures of each European country help account for either the widespread or the limited success of specific discoveries. Vaccines, such as tetanus shots, are required and easily available (in the United States, the latter is only administered in an emergency). In terms of aesthetic surgery, facial care is just as successful in Europe as in the United States, but breast enlargement has limited success.

Regionally, it remains interesting to discover the new uses of old technologies. Outdated computers being used as doorstops comes to mind, but on a more serious note, the acceleration of technological obsolescence has also introduced the dimension of nostalgia in everyday practices. In France and Germany, the ubiquitous fountain pen remains a standard requirement in schools, even though the erasable ballpoint has long been available (and is far less messy).

Tastes affect such adoptions in important ways, which anthropologists and sociologists investigate for marketers. The Volkswagen Golf, for example, had a lasting impact on youths in the 1980s, but its successor, notably the redesigned Beetle, has not caught on the way it did in the United States.

Other external factors that influence varying approaches to science and technology involve such matters as economic well-being and ecological concerns. In the 1970s, E. F. Schumacher published *Small Is Beautiful*, which involved a plea for adjusting technology to individual needs. While individuality is difficult to achieve in an economy of scale, the power of public lobbying and protest has definitely affected corporations.

OIL

The 1970s characterized the end of the European economic miracle, not the least because of the double oil shocks of 1973 and 1979. Both outcomes of events in the Middle East (the Yom Kippur War and the Iranian Revolution) and these oil traumas made governments (and populations) realize how vulnerable they were to economic pressures. In Germany, the Netherlands, and Switzerland, car-free days were introduced experimentally (they were eventually withdrawn due to reasons ranging from enforceability to rejected referendum proposals).

FEAR OF THE MACHINE . . .

A series of environmental disasters in the late twentieth century inspired considerable distrust in European public opinion and boosted the claims of ecological groups regarding the risks associated with large industrial endeavors. While the fear of nuclear power is one of the more obvious concerns, chemical accidents also grabbed headlines in Europe. In Flixborough, United Kingdom, on June 1, 1974, a chemical factory exploded due to a leak of cyclohexane (used to make nylon). Because the accident happened on a Saturday, 28 people died, but hundreds would have likely disappeared in the explosion had it been a workday.

Two years later, on July 10, 1976, a cloud of dioxin was released from a chemical plant in Seveso, Italy. Hundreds of acres of arable land were destroyed, and thousands of people suffered burns and breathing difficulties during the regional evacuation.

Both incidents, combined with tanker catastrophes off the Atlantic European coasts, such as that of the Torrey Canion (1967) or the Amoco Cadiz, such as (1978), instilled an awareness that industrial accidents no longer could be limited to their immediate surroundings.

... AND THE LOVE OF IT

In parallel, the concept of globalization has democratized the travel experience and made cheaper many technologies that once were considered luxury items. The variety of media and most recently the Internet have meant that Europeans are more aware, if they wish to be, of their worldwide surroundings. Thus, it becomes difficult in the present day to speak of a unique European experience when discussing science and technology. There is a seamless web beyond the realm of electronics that links Europe to other continents and blurs boundaries. On the other hand, *how* Europeans make use of science and technology may define their identity, from state regulation to everyday consumer items. Rather than destroy European identity, globalization will help Europe evolve into new directions.

ACROSS THE LAND, ACROSS THE SEA,
KIND MESSAGES I SEND TO THEE

Globalization around 1900: a British postcard pictures the impact of swift communication on the globe. Source: Author's collection.

BIBLIOGRAPHY

Adas, Michael. *Machines as the Measure of Men. Science, Technology, and Ideologies of Western Dominance.* Ithaca, NY: Cornell University Press, 1990.

Akrich, Madeleine, and Cécile Méadel, ed. *Energie, l'heure des choix.* Paris: Les Editions du Cercle d'Art, 1999.

"America against Europe: The Old World defends itself." *Suisse sportive* (March 1926): 25.

Ammon, C. E. "Spanish Flu Epidemic in 1918 in Geneva, Switzerland." *Euro-surveillance* vol. 7, 12 (December 2002): 190–92.

Auerbach, J. Λ. *The Great Exhibition of 1851: A Nation on Display.* New Haven, CT: Yale University Press, 1999.

BBC. "Survival Tips." *BBC Stop Look Listen.* March 11, 2007 http://news.bbc.co.uk/1/hi/magazine/4743384.stm.

Bennett, David. *Metro: The story of the Underground Railway.* London: Mitchell Beazley, 2004.

Bobrick, Benson. *Labyrinths of Iron. A History of the World's Subways.* New York: Newsweek Books, 1981.

Boch, Rudolf, ed. *Geschichte und Zukunft der deuitschen Automobilindustrie.* Stuttgart: Franz Steiner, 2001.

Borgé, Jacques, and Nicolas Viasnoff. *Archives des cheminots.* Paris: Michèle Trinckvel, 1995.

Borneque, E. M. "Comment l'Amérique conçoit la route." *Science et Vie* 344 (1946): 195–205.

Bridenthal, Renate, Atina Grossman, and Marion Kaplan, eds. *When Biology Became Destiny. Women in the Weimar Republic and Nazi Germany.* New York: Monthly Review Press, 1984.

Buchanan, R. A. *The Power of the Machine: The Impact of Technology from 1700 to the Present.* New York: Viking, 1992.

Cavelier, Patrice, and Olivier Morel-Maroger. *La Radio.* Paris: PUF, 2005.

Challe, Odile. "Le Minitel: la télématique à la française." *The French Review* 62, 5 (1989): 843–56.

Chant, Colin. *The European Cities & Technology Reader: Industrial to Post-Industrial City.* London: Routledge, 1999.

Chew, V. K. *Talking Machines.* London: HMSO, 1967.

Coffin, Judith G. "Credit, Consumption, and Images of Women's Desires: Selling the Sewing Machine in Late Nineteenth-Century France." *French Historical Studies* 18, 3 (Spring, 1994): 749–83.

Cooper, Howard. "Lorries and Lorry Driving in Britain, 1948–1968: The End of an Era." *Journal of Popular Culture* 29, 4 (1996): 69–81.

Coopersmith, Jonathan. *The Electrification of Russia, 1880–1926.* Ithaca, NY: Cornell University Press, 1992.

Courtney, Chris. "Miracle of Bern." May 22, 2007. http://www.soccertimes. com/oped/2006/courtney/jun28.htm.

Cowan, Ruth Schartz. "The Industrial Revolution in the Home: Household Technology and Social Chang in the Twentieth- Century." *Technology and Culture* 17, 1 (January 1976): 1–23.

Cramois, Pierre. "Les nouveaux telephones." *Mondes et voyages* 30 (March 15, 1932): 171.

Crane, Frank. "La voiture et la civilization." *Journal de Genève* March 26, 1923.

Crosland, Maurice. *The Society of Arcueil: A View of French Science at the Time of Napoleon I.* Cambridge, MA: Harvard University Press, 1967.

Cunningham, Hugh. *Children and Childhood in Western Society.* New York: Longman, 1995.

Davies, Robert B. "Peacefully Working to Conquer the World: The Singer Manufacturing Company in Foreign Markets, 1854–1889." *The Business History Review* 43 (Autumn, 1969): 299–325.

"Der Kaiser und das Motorwesen," *Motor* (February 1913): 25.

des Cars, Jean. *L'Orient Express: Un siècle d'aventures ferroviaires.* Paris: Denoël, 1984.

de Sola Pool, I., ed. *The Social Impact of the Telephone.* Cambridge, MA: MIT Press, 1977.

de Syon, Guillaume. *Zeppelin! Germany and the Airship, 1900–1939.* Baltimore, MD: The Johns Hopkins University Press, 2002.

de Syon, Joëlle, and Brigitte Sion. *Un siècle de progrès automobile. Salon international de Genève, 1905–2005.* Geneva: Slatkine, 2004.

Edwards, Clive. "Technology Transfer and the British Furniture Making Industry, 1945–1955." *Comparative Technology Transfer and Society* vol. 2, 1 (April 2004): 71–98.

Empeyta, Charles-Louis. *Bulletin de l'ACS* (1907): 3.

Fédorovski, Vladimir. *Le roman de l'Orient Express*. Monaco: Éditions du Rocher, 2007.

Fisher, Kate. *Birth Control, Sex, and Marriage in Britain, 1918–1960*. Oxford: Oxford University Press, 2006.

Fiske, John. *Television Culture*. London: Methuen, 1987.

Fletcher, Amy L. "France Enters the Information Age: A Political History of Minitel." *History and Technology* 18, 2 (2002): 103–19.

Forty, Adrian. *Objects of Desire*. New York: Pantheon, 1986.

Freeman, Michael, and Derek Aldcroft, eds. *Transport in Victorian Britain*. Manchester: Manchester University Press, 1985.

Friedlander, Henry. *The Origins of Nazi Genocide. From Euthanasia to the Final Solution*. Chapel Hill: University of North Carolina Press, 1995.

Frost, Robert L. *Alternating Currents: Nationalized Power in France, 1946–1970*. Ithaca, NY: Cornell University Press, 1991.

Frost, Robert L. "Machine Liberation: Inventing Housewives and Home Appliances in Interwar France." *French Historical Studies* 18 (Spring 1993): 109–30.

Galbi, Douglas. "Child Labor and the Division of Labor in the Early English Cotton Mills." *Journal of Population Economics* 10 (1997): 357–75.

Gardey, Delphine. "Mechanizing Writing and Photographing the Word: Utopias, Office Work, and Histories of Gender and Technology." *History and Technology* 17, 4 (2001): 319–52.

Garland, Ken. *Mr. Beck's Underground Map*. London: Capital Transport Publishing, 1994.

Gaskell, P. *The Manufacturing Population of England* [London, 1833]. May 22, 2007. http://www.victorianweb.org/history/workers2.html.

Gideon, Siegfried. *Mechanization Takes Command: A Contribution to Anonymous History*. New York: W. W. Norton, 1969 [1948]

Gillispie, Charles. "The Scientific Importance of Napoleon's Egyptian Campaign." *Scientific American* (November 1994): 78–85.

Goetz, Alisa, ed. *Up, Down, Across: Elevators, Escalators and Moving Sidewalks*. London: Merrell, 2003.

Goggin, Gerard. *Cell Phone Culture: Mobile Technology in Everyday Life*. London: Routledge, 2006.

Goodman, David. *The European Cities and Technology Reader*. New York: Routledge, 1999.

Hamre, Louis. "Dans les entrailles d'un steamer." *Mondes et voyages* 8 (15 April 1931): 250–51.

Hannah, Leslie. *Electricity Before Nationalization. A Study of the Development of Electricity Supply in Britain to 1948*. Baltimore, MD: The Johns Hopkins University Press, 1979.

Hannoun, Claude. *La vaccination*. Paris: PUF, 1999.

Hård, Michael, and Andrew Jamison. *Hubris and Hybrids. A Cultural History of Technology and Science.* London: Routledge, 2005.

Hård, Michael, and Andrew Jamison, ed. *The Intellectual Appropriation of Technology. Discourses on Modernity, 1900–1939.* Cambridge, MA: MIT Press, 1998.

Harp, Stephen L. *Marketing Michelin. Advertising and Cultural Identity in Twentieth-Century France.* Baltimore, MD: The Johns Hopkins University Press, 2001.

Hecht, Gabrielle. *The Radiance of France. Nuclear Power and National Identity after World War II.* Cambridge, MA: The MIT Press, 2000.

Hentschel, Klaus, and Ann Hentschel, eds. *Physics and national Socialism: An Anthology of Sources.* Basel: Birkhäuser, 1996.

Herlihy, David V. *Bicycle: The History.* New Haven, CT: Yale University Press, 2004.

Hessler, Martina. *Mrs. Modern Woman. Zur Sozial- und Kulturgeschichte der Hauhalttechnisierung.* Frankfurt am Main: Campus, 2001.

Holliday, Laura Scott. "Kitchen Technologies: Promises and Alibis, 1944–1966." *Camera obscura* 47, 16, 2 (2001): 79–131.

Hooker, Sir Stanley. *Not Much of an Engineer.* London: Crowood, 1991.

Hughes, Thomas P. *Networks of Power: Electrification in Western Society, 1880–1930.* Baltimore, MD: The Johns Hopkins University Press, 1983.

Hutchby, Ian, and Jo Moran-Ellis, eds. *Childhood, Technology and Culture: The Impacts of Technologies in Children's Everyday Lives,* London: RoutledgeFalmer, 2001.

Just, Willy. "Letter to SS-Obersturmbannfuhrer Walter Rauff, June 5, 1942." *Nazism: A History in Documents and Eye Witness Accounts, 1941–1945,* vol. 2, document 913. January 15, 2007. http://www.nizkor.org/ftp.cgi/people/r/rauff.walter/rauff.letter.060542.

Kern, Stephen. *The Culture of Time and Space, 1880–1918.* Cambridge, MA: Harvard University Press, 1983.

Kicherer, Sibylle. *Olivetti. A Study of the Corporate Management of Design.* New York: Rizzoli, 1990.

Kirchner, Wilhelm. "Das Geschäftsautomobil," *Motor* (May 1914): 60.

Kirp, David L., and Ronald Bayer, ed. *AIDS in the Industrialized Democracies.* New Brunswick, NJ: Rutgers University Press, 1992.

König, Wolfgang. *Volkswagen, Volksempfänger, Volksgemeinschaft. "Volksprodukte" im Dritten Reich: Vom Scheitern einer nationalsozialistischen Konsumgesellschaft.* Paderborn: Schöningh, 2004.

Kovári, K., and R. Fechtig. *Percements historiques de tunnels alpins en Suisse.* Zurich: Société pour l'art de l'ingénieur civil, 2000.

Kubicek, Robert V. "The Design of Shallow-Draft Steamers for the British Empire, 1868–1906." *Technology and Culture* 31, 3 (1990): 427–50.

Kühl, Stefan. *The Nazi Connection: Eugenics, American Racism, and German National Socialism.* New York: Oxford University Press, 1993.

Kupper, Patrick. *Atomenergie und gespaltene Gesellschaft. Die Geschichte des gescheiterten Projektes Kernkraftwerk Kaiseraugust.* Zurich: Chronos Verlag, 2003.

Landes, David. S. *The Unbound Prometheus. Technological Change and Industrial Development in Western Europe from 1750 to the Present.* New York: Cambridge University Press, 1969.

Landman, J. H. "Race Betterment by Human Sterilization," *Scientific American* (June 1934), 292–95.

Lenief, D. "Le VIIIe salon des arts ménagers." *Mondes et voyages* 4 (February 15, 1931): 106–7.

"Les tribulations de l'Orient Express." *Historia* 8–9 (October-November 1999): 43–49.

Loewy, Raymond. *Never Leave Well Enough Alone.* Baltimore, MD: The Johns Hopkins University Press, 2002 [1951].

Loudon, Irvine. *Western Medicine. An Illustrated History.* New York: Oxford University Press, 1997.

Lundin, Per. "American Numbers Copied! Shaping the Swedish Postwar Car Society." *Comparative Technology Transfer and Society* 2, 3 (December 2004): 303–34.

Maddison, Angus. *L'économie mondiale, Statistiques,* Paris: OECD, 2003.

Massard-Guilbaud, Geneviève. "La régulation des nuisances industrielles urbaines (1800–1940)." *Vingtième Siècle. Revue d'histoire* 64 (October 1999): 53–65.

McKay, John P. *Tramways and Trolleys. The Rise of Urban Mass Transport in Europe.* Princeton, NJ: Princeton University Press, 1976.

Méchin, Jeanne, and André Méchin. "Voyage à bord du Paul Lecat entre Marseille et Singapour en Novembre 1923." January 11, 2008. http://www.es-conseil.fr/pramona/vqpaq.htm.

Mohun, Arwen P. *Steam Laundries. Gender Technology, and Work in the United States and Great Britain, 1880–1940.* Baltimore, MD: The Johns Hopkins University Press, 1999.

"Monaco et sa course." *L'illustration* (May 1929): 7.

Moran, Joe. "Crossing the Road in Britain, 1931–1976." *The Historical Journal* 49, 2 (2006): 477–96.

Morton, Alan Q. *Science in the 18th Century.* London: Science Museum, 1993.

Möser, Kurt. *Geschichte des Autos.* Frankfurt: Campus, 2002.

Neft, David. "Some Aspects of Rail Commuting: New York, London, and Paris." *The Geographical Review* 49, 2 (1959): 151–63.

O'Connell, Sean. *The Car in British Society: Class, Gender and Motoring, 1896–1939.* Manchester: Manchester University Press, 1998.

Orland, Barbara. "Turbo-Cows: Producing a Competitive Animal in the Nineteenth and Early Twentieth Centuries." *Industrialising Organisms.*

Ed. Susan R. Schrepfer and Philip Scranton. New York: Routledge, 2003: 167–89.

Pearce, William L. "Science, Education and Napoleon I." *Isis* 47, 4 (1956): 369–82.

Pitkin, Donald D. *The House That Giacommo Built. History of an Italian Family 1898–1979.* New York: Cambridge University Press, 1985.

"Playing Safe on the Internet." March 16, 2008. http://ec.europa.eu/justice_home/funding/2004_2007/daphne/illustrative_cases/illustrative_cases_en/02_07_internet_action_en.pdf.

Potts, Malcolm, and Martha Campbell. "History of Contraception." *Gynecology and Obstetrics* 6, 8 (2000).

Prost, Antoine. "The Transition from Neighborhood to Metropolis." *A History of Private Life. V. Riddles of Identity in Modern Times.* Ed. Antoine Prost and Gérard Vincent. Cambridge, MA: Harvard University Press, 1991: 103–43.

Purcell, Carroll. "'Am I a Lady or an Engineer?' The Origins of the Women's Engineering Society in Britain, 1918–1940." *Technology and Culture* 34, 1 (1993): 78–97.

Quataert, Jean H. "The Shaping of Women's Work in Manufacturing: Guilds, Household, and the State in Central Europe, 1648–1870." *American Historical Review* 90, 5 (December 1985): 1122–48.

Razac, Olivier. *Barbed Wire. A Political History.* New York: The New Press, 2002.

Renneberg, Monika, and Mark Walker. *Science, Technology and National Socialism.* New York: Cambridge University Press, 1994.

Rouaud, Jacques. *60 ans d'arts ménagers.* Paris: Syros, 1993.

Royal Naval Museum. "The Navy League." April 2, 2008. http://www.royalnavalmuseum.org/info_sheets_navy_league.htm.

Rüger, Jan. *The Great Naval Game: Britain and Germany in the Age of Empire.* New York: Cambridge University Press, 2007.

Sachs, Wolfgang. *For Love of the Automobile. Looking Back into the History of Our Desires.* Berkeley: University of California Press, 1992.

Saldern, Adelheid von. "Cultural Conflicts, Popular Mass Culture, and the Question of Nazi Success: The Eilenriede Motorcycle Races, 1924–39." *German Studies Review* 15, 2 (May 1992): 317–38.

Saraceno, Chiara. "The Italian Family: Paradoxes of Privacy." *A History of Private Life. V. Riddles of Identity in Modern Times.* Ed. Antoine Prost and Gérard Vincent. Cambridge, MA: Harvard University Press, 1991: 451–503.

Schivelbusch, Wolfgang. *The Railway Journey: The Industrialization and Perception of Time and Space.* Berkeley: University of California Press, 1987.

Showstack Sassoon, Ann, ed. *Women and the State.* London: Hutchinson, 1987.

Smil, Vaclav. *Creating the Twentieth Century. Technical Innovations of 1867–1914 and Their Lasting Impact*. New York: Oxford University Press, 2005.

Smil, Vaclav. *Enriching the Earth: Fritz Haber, Carl Bosch and the Transformation of World Food Production*. Cambridge, MA: MIT Press, 2001.

Spreadbury, Benjamin. "The Milking Machine." *Rural History Today* 4 (January 2003): 6.

Standage, Tom. *The Victorian Internet*. New York: Berkley, 1999.

Strakosh, George R. *Vertical Transportation: Elevators and Escalators*. New York: John Wiley, 1967.

Thalheim, Matthias. "Ein Zeitstrahl durch 80 Jahre Hörspiel-Geschichte." May 2007. http://www.mdr.de/mdr-figaro/hoerspiel/1306615.html.

Thompson, Christopher S. *The Tour de France: A Cultural History*. Berkeley: University of California Press, 2006.

Trebilcock, Clive. "Industrialization of Modern Europe." *Oxford Illustrated History of Modern Europe*. Ed. T.C.W. Blanning. New York: Oxford University Press, 1996.

Treichler, Hans Peter. *La Suisse au tournant du siècle*. Zürich: Selection, 1985.

Vallgårda, Signild. "Problematizations and Path Dependency: HIV/AIDS Policies in Denmark and Sweden." *Medical History* 51, 1 (January 2007): 99–112.

Van Bath, B. H. Slicher. *The Agrarian History of Western Europe A.D. 500–1850*. New York: St. Martin's, 1963.

Vigarello, Georges. *Le propre et le sale*. Paris: le Seuil, 1987.

Villermé, Louis. *Tableau de l'état physique et moral des ouvriers employés dans les manufactures de cotton, de laine et de soie*. Paris: Renouard, 1840.

Volti, Rudi. *Cars and Culture. The Life Story of a Technology*. Baltimore, MD: The Johns Hopkins University Press, 2006 [2004].

Wall, Robert. *Ocean Liners*. New York: E. P. Dutton, 1977.

Wear, Andrew, ed. *Medicine in Society*. New York: Cambridge University Press, 1992.

Weber, Eugen. *Peasants into Frenchmen. The Modernization of Rural France, 1870–1914*. Stanford, CA: Stanford University Press, 1976.

Williams, Robert C. *Artists in Revolution. Portraits of the Russian Avant-Garde, 1905–1925*. Bloomington: University of Indiana Press, 1977.

Wohl, Robert. *A Passion for Wings*. New Haven, CT: Yale University Press, 1995.

Wohl, Robert. *The Spectacle of Flight: Aviation and the Western Imagination, 1920–1950*. New Haven, CT: Yale University Press, 2005.

Woodman, Richard. *The History of the Ship*. Guilford, CT: Globe Pequot, 2002.

Wormbs, Nina, and Arne Jernelöv. "You Might as Well Go Home When the System Is Down." *Framtider* 2 (2005): 25–30.

Yarwood, Doreen. *Five Hundred Years of Technology in the Home*. London: B. T. Batsford, 1983.

Zeitlin, J. "Flexibility and Mass Production at War: Aircraft Manufacture in Britain, the United States, and Germany, 1939–1945." *Technology and Culture 36*, 1 (1995): 46–79.

Zeitlin, J., and G. Herrigel. *Americanization and Its Limits.* Oxford: Oxford University Press, 2000.

Zeller, Thomas. *Driving Germany: The Landscape of the German Autobahn, 1930–1970.* Providence, RI: Berghahn Books, 2007.

INDEX

About the Author

GUILLAUME DE SYON teaches history at Albright College. He is the author of *Zeppelin! Germany and the Airship, 1900–1939,* and of articles on the social and cultural history of technology.